CAMBRIDGE LIBRARY COLLECTION

Books of enduring scholarly value

Life Sciences

Until the nineteenth century, the various subjects now known as the life sciences were regarded either as arcane studies which had little impact on ordinary daily life, or as a genteel hobby for the leisured classes. The increasing academic rigour and systematisation brought to the study of botany, zoology and other disciplines, and their adoption in university curricula, are reflected in the books reissued in this series.

Memorials of Sir C.J.F. Bunbury

Sir Charles James Fox Bunbury (1809–86), the distinguished botanist and geologist, corresponded regularly with Lyell, Horner, Darwin and Hooker among others, and helped them in identifying botanical fossils. He was active in the scientific societies of his time, becoming a Fellow of the Royal Society in 1851. This nine-volume edition of his letters and diaries was published privately by his wife Frances Horner and her sister Katherine Lyell between 1890 and 1893. His copious journal and letters give an unparalleled view of the scientific and cultural society of Victorian England, and of the impact of Darwin's theories on his contemporaries. Volume 2 covers the years 1844–8, and shows how Bunbury's marriage brought him into close contact with the geologist Charles Lyell, whom he greatly admired, and who became his brother-in-law. His diaries mention attending lectures by Sedgwick and Owen, and socialising with Babbage and Henslow among others.

T0188076

Cambridge University Press has long been a pioneer in the reissuing of out-of-print titles from its own backlist, producing digital reprints of books that are still sought after by scholars and students but could not be reprinted economically using traditional technology. The Cambridge Library Collection extends this activity to a wider range of books which are still of importance to researchers and professionals, either for the source material they contain, or as landmarks in the history of their academic discipline.

Drawing from the world-renowned collections in the Cambridge University Library, and guided by the advice of experts in each subject area, Cambridge University Press is using state-of-the-art scanning machines in its own Printing House to capture the content of each book selected for inclusion. The files are processed to give a consistently clear, crisp image, and the books finished to the high quality standard for which the Press is recognised around the world. The latest print-on-demand technology ensures that the books will remain available indefinitely, and that orders for single or multiple copies can quickly be supplied.

The Cambridge Library Collection will bring back to life books of enduring scholarly value (including out-of-copyright works originally issued by other publishers) across a wide range of disciplines in the humanities and social sciences and in science and technology.

Memorials of Sir C.J.F. Bunbury

VOLUME 2: MIDDLE LIFE PART 1

EDITED BY
FRANCES HORNER BUNBURY
AND KATHARINE HORNER LYELL

CAMBRIDGE
UNIVERSITY PRESS

CAMBRIDGE UNIVERSITY PRESS

Cambridge, New York, Melbourne, Madrid, Cape Town,
Singapore, São Paolo, Delhi, Tokyo, Mexico City

Published in the United States of America by Cambridge University Press, New York

www.cambridge.org
Information on this title: www.cambridge.org/9781108041133

© in this compilation Cambridge University Press 2011

This edition first published 1890
This digitally printed version 2011

ISBN 978-1-108-04113-3 Paperback

MEMORIALS

OF

Sir C. J. F. Bunbury, Bart.

EDITED BY HIS WIFE.

THE SCIENTIFIC **PARTS OF** THE WORK REVISED BY
HER SISTER, MRS. LYELL.

MIDDLE LIFE.

Vol. I.

MILDENHALL:
PRINTED BY S. R. SIMPSON, MILL STREET.
MDCCCXC.

Charles James Fox Bunbury,

WAS MARRIED ON THE 30TH MAY, 1844,

TO

FRANCES JOANNA,

SECOND DAUGHTER OF LEONARD HORNER.

AFTER A TOUR IN SCOTLAND,

AND PAYING A FEW VISITS, THEY ARRIVED IN THE

MIDDLE OF AUGUST, AT

MILDENHALL, SUFFOLK,

ONE OF THE ESTATES OF HIS FATHER,

SIR HENRY EDWARD BUNBURY, BART.

LETTERS.

CHAPTER I.

To Mrs. Charles Lyell.

Mildenhall,
August 20th, 1844.

My Dear Mary,   1844.

We have been now three days in our new
habitation, but I still feel rather like Robinson
Crusoe when first cast ashore, or perhaps, more
like a fish out of water; and yet I am in reality
very well off, having a good weather-tight house
(large enough and to spare), plenty of furniture
and books, and above all a most superlatively
good wife. But I must try to give you some
slight idea of our abode; it is an old irregularly
shaped house, all ins and outs and odd gables
and corners, having on two sides of it (the E. and
S.) a smooth green lawn, bounded by a thick belt
of dark trees and shrubs, beyond which is a wall
high enough to prevent the ignoble vulgar from
peeping at us. On the W. are the offices, and

B

1844. a large melancholy looking walled garden, which
my father has lately sold; on the N. the road;
but the part of the house looking towards this is
effectually cut off from our part by a thick
partition wall just built, so we are quite secluded,
and have no annoyance from the village or the
road. Our rooms on the ground floor are—first a
neat moderate sized dining room, looking to the
East, next the drawing room, a very cheerful
and pleasant room of an irregular form, with
three windows to the E. and a bow window to
the South, opening down to the ground; thirdly
the book room, in which I am now writing, and in
which I have been very busy these last two mornings
arranging my books on the shelves. This also is
a cheerful and pleasant room. A large hornbeam
stands at the S.E. angle of the house, close to the
drawing room windows, and a fine acacia next to
the window of this room. I need not enter into
the detail of the bedrooms, but they are sufficient
to allow of our lodging several friends. At the
top of the house is an old, queer, gloomy, long
gallery, and at one end of it a large room which
is to be my museum; and adjoining this is a
small room which we have christened *Sky parlour*,
in remembrance of Kinnordy. Such is our den in
which we shall be delighted to welcome you
and Lyell, if, as I hope, you will pay us a visit
before you go to York. We shall be able to lodge
you comfortably, and I hope we shall be able to
give you *something* to eat, though not to feast you.

I forgot to mention that within our boundary-wall

besides the lawn and shrubbery, we have a paddock 1844. with some fine trees in it, a kitchen garden in a very rubbishy condition, and a small green-house, which is just going to tumble down. Over the said wall is seen the tower of the Church, a large and fine Church, in which we have a most dismal looking dilapidated pew, big enough to hold a company of soldiers, and so far from the parson that we can hardly hear a word. Did you ever read Tennyson's poem of " Mariana in the Moated Grange?" I think Mildenhall would put you rather in mind of it.

Dear Fanny takes most zealously to her new occupations, and encounters the plagues of house-keeping with a most gallant spirit, but she sometimes bewilders me not a little by asking questions about household affairs, of which I know about as much as a pig knows of pure mathematics. She is every-thing that is affectionate and good and charming, but so far from strong that she cannot walk half a mile, which is a sad pity ; but she consoles me by saying that she feels well in other respects. I was very happy to find that my father and step-mother liked her so much when they became really acquainted with her. They have been most kind in getting this house in order for us and providing us with everything to make us comfortable, in our new way of life, and I am sure I may consider myself in every way uncommonly fortunate. I am very happy to find by your letter which Fan has received this morning (21st)

1844. that you do mean to pay us a visit in September.

Pray give my love to mamma, &c., and let them see this letter if they want to know all about Mildenhall.

Your very affectionate brother-in-law,

C. J. F. BUNBURY.

JOURNAL.

The 10th day since my beloved wife and I took 1844.
up our residence at Mildenhall, and entered on
a course of life new to us both. We are beginning
by degrees to feel ourselves more settled and
at home, but are yet far from being completely
so.

Since we began our residence here, I have read
through rather more than half the first volume
of the Memoirs of Francis Horner, which I find
interesting, though not light reading. Yesterday
morning I began the Odyssey, and read 105 lines
of it.

Went on with the Odyssey, l. 106—212. I do
not as yet find it interesting.

Went with Curling* into Burnt Fen to observe

* His Father's agent.

1844. the operation of claying, as it is called, by which
the land in that district has been improved to a
remarkable degree and rendered capable of producing very rich crops. Holes are dug down
through the black peaty soil to the clay (or rather
marl), which lies beneath it; and this is thrown
up, and thrown over the surface of the soil, becoming gradually mixed with the soil by the action
of the weather, and especially the frosts and thaws
of winter. The holes are filled up again with
the peat. The "clay" which is thus used seems
to be more properly marl, as it contains much
calcareous matter ; it is white when dry, moderately
unctuous when moist, contains few shells, but is
much mixed with decayed vegetable matter. It
is probably a very recent deposit. I could not
learn that they have ever penetrated through it,
to learn what formation lies under it. The peat
which covers it, in the part where I saw the men
at work, is about eight or nine feet deep. The
beneficial action of this clay on the peaty soil,
is supposed to consist in giving it more consistence
and tenacity, and perhaps the calcareous matter
is also rather advantageous.

This claying process, united with effectual drainage, has produced a remarkable change in the value
of the land in Burnt Fen. Land which was
formerly mere swamp is now covered with fine crops
of wheat, and affords a rent of 25/- the acre. The
same rotation of crops is practised on this reclaimed
fen as on the higher lands.

This district now retains little of the appearance 1844. of a fen, except in the broad wet ditches which still afford a few marsh plants.

Burnt Fen belongs to a separate system of drainage from Mildenhall Fen, and is much more effectually drained, having for some years had the advantage of a steam engine. A part of it is in Cambridgeshire.

The digging of holes to extract the clay is very hard work, and is well paid. A good labourer may earn 3/- to 4/- a day at it.

The parish of Mildenhall is of uncommon extent; from the Town to the boundary of the parish (which is also the boundary of the county), in the direction we went this day, viz. N. W., it is very nearly seven miles. The number of acres in the parish is about 16,000; the population by the last census upwards of 3,700.

September 3rd.

Finished the first volume of the Life of Francis Horner.

September 22nd.

Since I wrote the last entry in this Journal, our seclusion has been very agreeably interrupted by visits from some of our dearest friends. Charles and

1844. Mary Lyell stayed with us from the 7th to the 11th, when we went with them to Barton, and met there my wife's father and mother and her sister Katharine. With these latter we returned home on the 14th, and they remained with us till the 19th.

I cannot put down any positively new fact, or very striking remark, that I heard during this time ; yet I am sure that I am improved both intellectually and morally by such society. My regard and admiration for Charles Lyell increases as I know him more intimately. He has one of the most happily constituted minds that I am acquainted with ; a remarkable degree of energy, activity and ardour, united with the utmost clearness and soundness of judgment ; strong and steady affections, high principles, great decision and firmness of purpose, and a knowledge of men and of society, which I should think is not often found united with such enthusiasm in the pursuit of science.

I feel strongly his superiority to me in all respects, yet the superiority is the reverse of oppressive or discouraging ; I find that his conversation and example have always a most salutary and invigorating effect on my mind. He has strongly urged me to take up the study of fossil botany, which he says is not now pursued with earnestness by anyone in this country.

Mildenhall,
Sept. 22nd.

Since our friends left us, I have returned to my

regular reading, and finding the first books of 1844. the Odyssey very dull, have made a fresh start with the 5th, which brings us to the cave of Calypso.

My scheme of study for this autumn and winter, is as follows:— To study fossil botany as far as it is practicable at a distance from any large collection; to keep up my knowledge of Greek, reading Homer and Xenephon, and perhaps Aristophanes; to revive my knowledge of Algebra and, if I can, of the Differential Calculus; and to learn German. I wish also, by degrees to make myself well acquainted with this estate, and with the people on it, to understand the management of it, and to attend to the condition of the poor; but this will be more out of door work. I will not neglect to keep up my interest in public affairs, and to strengthen my mind by studying the lives and opinions of great and good public men such as Horner and Romily. *Deus adjunct.*

23rd September.

The last portion of my collections arrived from Barton, and a considerable part of the morning was spent in unpacking and settling them.

Afterwards I read one hundred lines of the Odyssey, v. 96-202 of the fifth book.

October 3rd.

Employed myself this morning with a book which I borrowed from Curling, containing the names of all the tenants on this estate, with the quantity of land held, and the rent paid by each.

October 4th.

Went on from 11 to 12 o'clock, with my employment of the preceding day, writing out a classification of the tenants; then went to the petty sessions where I was detained till half-past two, in the most tedious, unsatisfactory and unprofitable routine work of trumpery cases.

Walked out after luncheon with my Fanny, and began to give her some instructions in muscology.

In the evening, began to study Adolphe Brongniart's *Végétaux Fossiles*.

October 5th.

At the Board of Guardians it is satisfactory to see how very large a proportion of the relief administered in this union is out-door relief. In the last quarter, the number of poor of this parish of Mildenhall, relieved in the *Workhouse*, was 16, viz:—5 men, 3 women, 8 children. The number receiving *out-door relief* was 76 men, 129 women, 169 children.

The number of paupers at present in the house is 36, viz:—5 men, 6 women and 25 children.

Of the 5 men, 4 are sick, and the other is a very old man, who stays in the Workhouse by choice, having been offered out-door relief. The total numbers receiving parish relief in this union of thirteen parishes, amounts to 127 men, 252 women, and 263 children.

From 12 o'clock to 2, I had a good spell of work at Adolphe Brongniart.

After luncheon took a walk with my dear wife, who is improving much in strength and power of walking.

October 19th.

At the Bench of Magistrates yesterday, when applications for exemption from payment of poor rates were heard ; there came before us some flagrant instance of the exorbitant rate at which the poor are often obliged to hire cottages. James Bonnett, of Lakenheath, having several children, and earning (I think) thirteen shillings a week, pays eight guineas a year for a small cottage, with only one bedroom, and with no land attached to it. In several other instances in the same parish, it appeared that six and seven guineas a year were charged for cottages of a very mean kind. In this parish, Mr. James Morley, who is the proprietor of many cottages, lets them generally at rents as high as five pounds or five guineas, whilst most of those belonging to my father are let at three guineas and under. In one instance Mr. Morley hired two cottages from my father for three guineas a year, and let them again to laborers at five pounds.

LETTERS.

To Mrs. Charles Lyell.

1844. My Dear Mary,

I understand that to-morrow will be your birthday, so I write to wish you many very happy returns of it; and heartily do I wish that you and your husband may very long continue to enjoy that happiness which you both so well deserve. Since my aunt and cousin* left us, we have been settling into pretty steady and satisfactory habits of work, but we shall be interrupted again to-morrow, as we are going to Barton to stay till after the Bury ball. I am very glad however that my dear Fanny should have some amusement. Moreover, as I am not a little proud of my wife, I am happy that she should appear to advantage among our neighbours, as I am sure she will.

I am deep in Adolphe Brongniart, and much interested and pleased with his great work on the whole, though I do not think it would be difficult

*His aunt, Lady Napier, and his youngest cousin, now Lady Aberdare.

to find matter for criticism. Fortunately the Ferns 1844
are a tribe of plants that have always been very
attractive to me, and I have a pretty large collection
of the recent species, though many of them un-
named. But I constantly regret that I so much
neglected the opportunities I had in Brazil of
studying the growth of the arborescent Ferns as well
as of Palms and so many other tribes of plants,
which one has no opportunity of studying in Europe.
I am most thankful to Lyell for his kindness in
confiding to me his American specimens, and for all
his assistance, and I hope I shall by-and-by be able
to be of use to him. I am sorry to hear you are
going to lose him for a week, which to Fanny and
me would seem a long separation.

I am going regularly through Adolphe Brongniart's
great work, comparing his descriptions and plates and
copying out all his generic and specific characters.

I am happy to say that my dear Fanny is
improving very much in strength and power of
walking, and is able to take really good walks with
me. We have been out several times on the warren
much further than she had ever been before, this
last week, and one day we walked as far as the little
fen, to which I took Lyell, the last day that you
were here, and where we found the Buckbean ; and
she was not at all the worse for her walk. She is
getting more colour in her cheeks, too, I think. She
has begun to study the mosses, and is a promising
pupil. In every respect she is a most excellent wife,
and I can never be sufficiently grateful for her
devoted attention to me ; her eagerness to serve and

1844. assist and encourage me in all my pursuits and objects, and her zeal to do good.

I do not like the spirit in which the *Times* commented on the meeting at Bury, and yet I cannot say, myself, that I have very high expectations from the proposed Society. Of the benefits of the allotment system, I have no doubt whatever (I mean when it is well administered and attended to). I most fully believe that my father has done very great good on his own estates, and that any gentlemen who will do on their own properties respectively what my father has done on *his*, will effect great good, but I have not very much faith in associations except where the object to be attained requires so great an expenditure as can only be met by the contributions of many; certainly I have not much faith in associations for doing collectively what each member might do separately. I have some fears that our Society may turn out merely a Society for *talking about* the allotment system.

With my kind love to Charles Lyell, and to those in Bedford Place, believe me,

Your very affectionate brother,

C. J. F. BUNBURY.

To Mr. Horner.

Mildenhall,
October 16th, 1844.

My dear Mr. Horner,

I was much obliged to you for your letter
of the 6th, and for the Manchester newspaper, in
which I read the speeches of young England with a
great deal of interest and pleasure. That party,
in spite of the ridicule with which it has been visited
seems to have a great deal that is good and valuable
in its aims and principles (mixed, no doubt, with a
good deal that is absurd), and I think it is certainly
rising, and ought to rise,in influence and importance.
It is a great pleasure to see factious enmities for
once laid aside, and men of all parties concurring
in a good purpose with so much apparent cordiality
as on the late occasion at Manchester. Even the
bitter spirit of the anti-corn-law league was softened
for a time into harmony with the opposite party.

I got Adolphe Brongniart's works from the
Geological Society, some time ago, and have been
studying them diligently, and with much interest.
The great *Histoire des Végétaux Fossiles* is a beautiful
book, and unquestionably of very high value.

I presume that the Memoir you allude to is one
on the internal structure of Sigillaria elegans, in the
Archives du Museum d'Hist. Nat. (1839), which I
have been reading very attentively, and of which I
have made a careful abstract, as it is of great interest

1844. and importance.   Adolphe Brongniart, with praise-
worthy candour, there gives up his former theory as
to the nature of Sigillariæ (viz. that they were tree
ferns) which in the *Histoire* he had supported with
most ingenious and elaborate arguments, and
comes to the conclusion that they were Dicotyledonous trees, more nearly allied to Zamia than to
anything else now existing.   I should be curious to
see Lindley and Hutton's Fossil Flora (which I
could not get from the Geological Society) though,
from what Henslow told me, I should suppose that
Lindley is much more hasty and less careful than
Brongniart. The subject of fossil botany is certainly
a very curious one, but appears to be more obscure
and difficult than any branch of the study of recent
plants, owing to the imperfect nature of most of the
materials we have to work with.   It is true however
that those materials are much less overwhelming in
point of quantity.

Lyell has been so good as to lend me all his
American collection of coal plants, the examination
of which, and the comparison of them with Brongniart's descriptions and figures, will be of very great
advantage to me. And if I can go to town in the
spring, I mean to devote myself to a diligent examination of the specimens in the Geological Society's
museum.   I am happy to say that my dear Fanny is
very well, and a good deal stronger and better able
to walk than at the time when you were here.   I am
beginning to read to her in the evenings General
Napier's History of the Peninsula War, not however

intending to go through the whole (which would 1844
be tedious), but to pick out the most interesting
parts. I have been so much occupied with fossil
botany, that I have not yet had time to begin
Dugald Stewart's book which you were so good as to
send me.

We are here suffering from a deluge, for it has
been raining excessively this last night and day, and
the house is very far from waterproof, so that in my
dressing room in particular, one might have *enjoyed*
a shower-bath without any trouble or expense. I
cannot help thinking how Susan would have
caricatured us if she had seen our disastrous
condition, with a little cascade pouring through the
ceiling of the drawing room, and everybody bringing
tubs and blankets and rugs to catch the water.

I hope your health is pretty good, and that you
are not much the worse for the hard work you have
had among the factories.

Pray give my love to Mamma and Katharine, and
believe me,

Your very affectionate son-in-law,

C. J. F. BUNBURY.

To Mr. Bunbury,
From his father, Sir Henry E. Bunbury, Bart.

Bath,
November 10th, 1844.

My dear Charles,

Emily and I have been passing some days at Bowood, with great pleasure. The house is so enjoyable, and everything is done with such quiet good taste, that one forgets its magnificence; but the grounds as well as the house itself, are beautiful. We had such a succession of excellent company, amongst whom are to be noticed, Hallam and Blake, Rogers and Moore, &c. We like Hallam very much, and as he is living at Clifton, we are likely to see a good deal of him this winter.

Emily and Cissy will probably remove to Clifton in the course of this opening week.

Pray send for Henry Gray, a labourer at West Row; question him about his allotment: tell him that his letter has been forwarded to me from Barton, and that I will not forget his application for more land. I wish very much to enlarge some of the allotments at M'hall, and to get some more pieces for labourers, as soon as Curling can persuade tenants to give up small portions of their farms for this purpose.

Next spring, I mean to build more cottages. There are not near enough belonging to me, in proportion to the size of the estate. Pray desire Curling to make out a list (against I return to Barton) of my cottages, and of the rent of each, exclusive of *allotments*, but including more gardens.

I hope you take the *Bury Post*. Henslow's letter 1844. dated October 27th, delights me. He has got on the right tack, and his boldness in opening that course to others deserves high praise. Emily is better for her visit to Bowood, but I shall be glad when she gets out of this cauldron Bath, and its murky atmosphere.

Much love to Fanny,

Ever affectionately yours,

H. E. B.

Such pictures at Bowood ! and such a library !

To Mrs. Charles Lyell.

Mildenhall,
November 26th, 1844.

My dear Mary,

Fanny and I are just come in from a constitutional walk to the chalk hill, beyond Barton Mills, where there is the large chalk pit to which Lyell and I walked one day, the air is cold, clear and bright, and the effects of light and the sunset colouring on the distant landscape, made it look almost pretty. I do not feel disinclined to take a nap after coming in, but I think *on the whole* that I shall be better employed in writing to you. It must be very pleasant to all the family party in Hart

1844. Street and Bedford Place, to find yourselves re-assembled, after such a dispersion.

We miss Leonora and Joanna, a good deal, and think with great pleasure of their visit, but we are very comfortable and happy in our secluded Eden, and the time passes away smoothly and swiftly. Fanny is most industrious and to my great satisfaction she has found out that her household affairs, school and old women, need not occupy all her time, but that she has time for good steady reading. She is deep in "Carlyle's Past and Present," and over head and ears in love with Abbot Sampson of St. Edmund's Bury, indeed to such a degree as actually to make me jealous, and I believe she wishes I were an Abbot! I am getting on with Lyell's coal plants, and am also busy with recent ferns and mosses, my love for the latter having been revived while Leonora was here, in consequence of my looking over my collections to pick out duplicates for her.

I am very glad that your father is likely to be President of the Geological Society.

It is time to dress, so I must say good bye, with love to all the family party.

Your very affectionate brother,
C. J. F. BUNBURY.

To Mr. Horner.

Mildenhall,
December 7th, 1844.

My Dear Mr. Horner,

I thank you much for your kind and 1844. interesting letter of the 1st. It gave me great pleasure to hear that you had been nominated President of the Geological Society, and that you had found yourself able to accept it ; and I promise myself much future pleasure in hearing your anniversary address. The nomination was an honour justly due to you as one of the founders (if I am not mistaken), and one of the earliest contributors to the publications of the Society, and I am very glad that your official chief did not think fit to throw any obstacle in your way. I am glad also that you have determined not to leave London at present, as I shall have a much better chance of seeing you frequently while you reside there, than if you removed into the North of England. I continue to take much interest in the study of fossil plants, and by studying Adolphe Brongniart and examining and comparing Lyell's North American specimens, I have made myself, I think, as familiarly acquainted with several of the plants of the Coal formation as I am with any recent species, so as to be sure of knowing them again wherever I meet with them. The Geological Society has been very liberal in allowing me to keep the volumes of Brongniart so long, and I hope I have made good use of them.

1844. Whatever is contained in those works, that is of a
more general nature, and not mere detail of specific
differences, I have not only studied, but carefully
abstracted, so as to have fully made myself master,
I think, of Brongniart's views on all the most
interesting questions connected with fossil botany,
(as far, I mean as they are contained in the volumes
I borrowed), and the Penny Cyclopædia supplies
me with the outlines of Lindley's opinions on the
same subject. Lindley is clever and ingenious but
hasty and crotchety. I think, and as Henslow said,
he seems to have proceeded on the one principle of
opposing Ad. Brongniart on *every* point where a
difference of opinion was possible. The latter is
inclined, I think like many other naturalists of the
present day, to multiply species on very slight
grounds; and perhaps he sometimes theorizes rather
rashly; but he has extraordinary merit in his treatment of the whole subject, and I particularly admire
the general dissertations which he prefixes to his
detailed account of each tribe, and the manner in
which he compares the fossil plants with those now
existing. With a view to such comparisons I have
continual reason to regret the scantiness of my
materials compared with those which I might have
collected in South America and in Africa. I mean
particularly in the way of specimens illustrating
the internal structure of stems, and other points of
vegetable anatomy and physiology. In fact, I feel
strongly, and with deep regret, how little use,
comparatively, I made of the leisure and great
opportunities I enjoyed in the former part of my life;

how much more I might have observed and learned, 1844.
and done towards the advancement of science; but
as it is useless to lament over opportunities which
are not likely to recur, I must only take care to
make better use of my time for the future. I have
no thoughts of abandoning the more attractive study
of recent plants, but I will certainly devote a large
share of my attention to the ancient Flora, as the
labourers in that field seem to be so much less
numerous, that one is more likely to be able to
contribute materially to the progress of science.

Henslow's articles in the Bury paper are excellent
in feeling and much to be commended for the
courage with which he attacks some of the worst evils
of our social state; but they are lengthily, heavily, and
confusedly written, and not likely, I am afraid, to
attract so much attention as they really deserve.
With regard to his idea of regulating the proportional
number of labourers to each farm, I hardly know
what to say. The farmers have so great an aversion
to the idea, that I apprehend it would hardly be
practicable, unless the dose were *sweetened* by a
reduction of rents; yet I much doubt whether, in the
long run, the farmers would be the worse for it. I
believe Henslow is perfectly right in saying that one
considerable source of evil in this part of the country,
is, that farmers are very apt to undertake more
land than they have either capital or skill to manage
well.

What you say on this subject is no doubt sound
political economy; but the conclusions of political
economy must be modified by moral considerations;

1844. and besides, it is surely a short-sighted policy to choose to maintain the labourers in the workhouse rather than pay them for useful labour. As for the abstract *right*, I conceive that a landlord has as much *right* to introduce into his lease a covenant regulating to a certain degree the amount of labour, as a covenant regulating the mode of cultivation; whether either restriction be expedient is a separate question. But the events here sufficiently show that no regulations respecting labour, however effectual for their immediate purpose, would prevent the crime of incendiarism. That diabolical practice still continues to be frequent in this part of the country, and (I am sorry to say), has revived in this parish, though there are no labourers out of employment and no peculiar distress. Distress in some degree is always likely to exist in a large, scattered and not wealthy parish like this; but it seems clear that the incendiaries are not to be looked for among those who really suffer the most. We have not yet, indeed, any clue to the individual criminal in the recent case, but there is little doubt that it was some one (or more) of a notorious set of men, who earn very good wages whenever they choose to work, but who are always drinking and making disturbances, and are so wasteful and profligate that their earnings never do them any gcod.

It is difficult to see any speedy remedy for this sort of evil. Education will doubtless do much ultimate good, but its effects cannot be speedy. But the whole subject is much too extensive and difficult to be satisfactorily discussed in a letter

of any reasonable length; and I am afraid this 1844.
letter is already one of *no* reasonable length.

I am happy to say that my dear Fanny is quite
well, notwithstanding the intense cold of the weather,
the like of which I think I never felt so early in
the winter. She is always busy and usefully employed, always cheerful, always the kindest and best
of wives, and I become continually more and more
sensible what a treasure I possess in her. We
found my father considerably the better for his stay
at Bath, and Lady Bunbury certainly in better
spirits, and therefore I should say in better health
than when she left Suffolk.

We have been reading Miss Martineau's wonderful histories with much amazement, and certainly
not with entire faith, though I do not in the least
suspect her of stating anything but what she
perfectly believes herself. But I will not spin out
this letter to a still more unconscionable length by
a dissertation on this curious subject.

Pray give my kind love to all your family in
Bedford Place and Hart Street, and believe me

Your affectionate son-in-law,
C. J. F. BUNBURY.

December 9th, 1844.

1844.

To Mrs. Horner.

Mildenhall,
December 20th.

My dear Mrs. Horner,

As to-morrow is your birthday, I write to wish you many happy returns of the day, and I am sure, that of your many friends, no one wishes it more cordially than I do. Independently of all the kindness you have shown me in other ways, I feel the warmest gratitude and affection towards you, as the mother of my dear and excellent wife, and I hope and trust you may long live to see your children happy around you, and to enjoy their love and gratitude. I wish we could join your happy family circle at this season, but though we cannot be present with you in the body, we are so in thought, and the delightful letters which Fanny frequently receives from you, and her sisters, quite place us by your fireside ; and make us participate in your amusements and occupations.

I am just come in from walking ; it has been a very fine, bright day, but a piercing cold wind. My darling wife is sitting by the fire reading Ariosto. She was anxious to go to her school to-day, but I dissuaded her from it, as I was sure the cold air would bring on her tooth-ache as bad as ever. My brother Henry dined and slept here yesterday ; he had been taking part in one of those great massacres called *battues*, in which he and five other gentlemen killed 600 head of game.

By the by, have you happened to see the Duke of 1844.
Grafton's pamphlet on game-preserving ?—it is very
unpalatable to our Suffolk squirearchy—but is full of
good feeling, and in my opinion, good sense too.   If
you or Mr. Horner would like to see it, I will send
you a copy.

I should like very much to be at the next anniversary of the Geological Society, when Mr. Horner
will, I suppose, enter upon his office.

I beg you to give my love to all your party, and
to believe me

Ever your affectionate son-in-law,
C. J. F. BUNBURY.

JOURNAL.

1845. My birthday. Received most kind letters and good wishes, and various pretty little presents from my dear wife's mother and sisters. How much a happier, and I hope, a wiser and a better man, I am than at this time last year. I am well aware how weak and faulty I still am, but I will do all I can to make myself more worthy of the love which my excellent wife bears to me, and that she may never repent having given herself to me.

I made out, by the help of a dictionary, two sentences of German in Göppert's Work on fossil ferns, being the first attempt I have ever made at reading that language.

Completed the examination and arrangement of my South American ferns of the groups Asplenieæ and Aspideæ. In the afternoon (the day being remarkably fine) walked to the hill marked in the ordnance map as Codson hill, and collected some

mosses, in particular, Jungermannia ciliaris, of which 1845.
however I found but a small quantity, after a
careful search.

In the evening, read some of Lamb's Essays
("Elia") to my Wife and Susan.

LETTERS.

To Mrs. Horner.

Mildenhall,
February 4th, 1845.

1845. My dear Mamma,

Many hearty thanks to you for your kind letter and your good wishes on my birthday, and also for the pretty present you have been so good as to send me, which I like very much. I am indebted likewise to my dear sisters for most kind notes and presents from each and all of them, but as I have not time to write to every one separately, I must beg you to give my hearty thanks and affectionate love to each and to all. I feel indeed very happy in enjoying the good will and esteem of my dear wife's excellent parents and sisters, and I think no man can be more fortunate in marriage (in every respect) than I am. I assure you I fully return the love which you all shew towards me, and to you in particular I feel most truly grateful for giving me such a treasure as my darling Fanny. Long may you live to enjoy the love and gratitude of your children, and to see them flourishing around you.

I enjoyed in the highest degree, Charles and Mary

Lyell's visit to us, only regretting that it was so 1845.
short ; and I am looking forward with (if possible)
still greater pleasure to our re-union in Bedford
Place on the 14th.

Susan's society is a great delight to both Fanny
and me; she has nearly got rid of her cold, and
we hope to produce her in fine condition when we
come to London. Fanny has almost completely
recovered her power of sleeping, and is, I think I
may say, very well, but she is rather overwhelmed
and worried by poor women, school business,
accounts and other domestic and parochial affairs,
and will, I am sure, be much better for a visit
to London, for change of scene and thoughts.

Again let me thank you, dear mamma, for your
very kind letter and token of remembrance ; and
pray assure my dear sisters (including Mary), that I
am very grateful to them also.

Believe me ever,

Your very affectionate son-in-law,

C. J. F. BUNBURY.

JOURNAL.

1845.   Occupied in the morning with examining under the microscope the beautiful structure of Jungermannia ciliaris, and comparing it with others of that genus; in the afternoon, writing letters. Very much entertained with two articles in the *Quarterly*, the one, by Mr. Hayward, on the " Harem " and the " Rights of Women," extremely lively and entertaining. The other on Kinglake's book called Eothen; in this case, the amusement not derived from the criticism, but from the extracts inserted.

Went to the Petty Sessions; no business. Worked again at Göppert's book, and made out a good bit of German, partly by the dictionary, partly by the help of my Fanny, who is very familiar with that language.

Hard at work again at Göppert's book, by which I am at once learning a language, and adding to my knowledge of fossil botany.

February 14th. 1845.

Arrived in London, and were most kindly welcomed by our dear friends in Bedford Place.

February 15th.

Went to the Geological Society with Mr. Horner,
who introduced me to Mr. Ansted, the vice-secretary,
and Mr. Woodward, the curator of the Museum.
A pleasant evening party at the Lyell's. I was
introduced to M. de Verneuil, a lively and agreeable man.

February 16th.

Went with Mr. and Mrs. Horner, Fanny, Susan,
etc., to the College of Surgeons, where we were
received by Owen, and saw its admirable Museum
to the greatest advantage under his guidance. Saw
the fine skeleton of the Mylodon, from La Plata; it
was most like a gigantic sloth, but seems to have
been formed particularly for uprooting trees,
not for hanging to their branches; its tail,
hind legs, and bones of the pelvis, are of
astonishing size and strength. To protect it from
the blows of falling trees, its brain was covered by
an additional bony plate, besides the usual skull,
in spite of which, this individual appears to have
had its skull fractured in two places. The vegetation
of that country, which at the present day is almost

D

1845. entirely herbaceous, must have been of a different
character when this animal existed; Owen said
that fine specimens of silicified wood were brought
from the same locality with these bones. The shell
or   cuirass   of   the   gigantic   extinct   Armadillo,
(Glyptodon),   also   from   La   Plata,   a   splendid
specimen.

Bones of the Dinornis, the great bird of New Zealand, making up a considerable portion of the skeleton,
have been lately set up, and by their side a skeleton
of the African Ostrich, for the purpose of comparison.
Owen says, the Dinornis was probably ten feet high,
while the Ostrich is from 6½ to 7; but the difference
in bulk was much greater than in height, the bones
of the Ostrich look very slender by the side of
those of the New Zealand bird. Owen thinks that
the statements of the New Zealanders relative to
this giant bird are not to be trusted, but are merely
traditions   analogous   to   those   of   the   American
Indians, relative to the Mastodon.   Among the
human skulls is one of a *Thug*, who was killed by a
blow on the head from one of his fellow prisoners;
the skull is almost as thin as an egg-shell, and was
broken literally *to pieces* by a blow of no great
violence.   The jaws project excessively.   Owen is
of opinion that this *prognathous* character, as Dr.
Pritchard calls it, is not really characteristic of any
particular race or races, but belongs to all persons of
whatever race, in whom the animal or sensual
qualities predominate greatly over the intellectual
and moral.

Worked for 3 hours at fossil plants in the Museum of the Geological Society; also looked through the first volume of Lindley and Hutton's Fossil Flora. A large evening party at home, Mr. Brown, Mr. Babbage, Mr. Hallam, Sir Edward Ryan, M. de Verneuil, Mr. Ansted, Mr. Forbes and others. I had a good deal of talk with Mr. Brown and Mr. Ansted. I asked Mr Brown if he had heard of the strange story, which has been much discussed in the *Gardener's Chronicle*, of the supposed transmutation of one sort of corn into another, of oats, I think, into rye. He would not express decidedly any opinion on the subject (according to his manner , but smiled sarcastically, and remarked that such a transmutation might be *very convenient*. He said, the author of " Vestiges of Creation" has availed himself of this supposed fact, imagining it to be favourable to his theory, to which it is in reality adverse ; for his theory is that of the *gradual* development of forms, whereas the fact in question, supposing it true, would be an instance of an alteration *per saltum*, from one form to another considerably different.

The same author has also introduced into his book Mr. Cross's supposed *creation* of insects, which Mr. Brown evidently disbelieves Mr. Brown told me that Psarolites have been found in Brazil, and that a very fine Psarolite in his possession, which was supposed to have come from Werner's collection, has been ascertained to be Brazillian. Mr. Ansted told me that Mr. Morris has lately been employed in

1845. describing some fossil plants, brought by Strelezki from the coal formation of New South Wales; that they are distinct indeed from those of Europe, but much less different than the recent vegetation of that country is from the European. This however, is not surprising, since closely-allied species of that tribe flourish at the present day in very remote countries.

———————

February 19th.

Worked again at coal plants in the Geological Society's Museum. Met Dr. Falconer at the G. S. He told me that in the coal mines of Burdwan, in Northern India, there are no Sigillarias, Stigmarias, nor Lepidodendrons; these tribes of plants, which prevail so much in the coal fields of Europe and North America, appear to be entirely absent from that Indian coal field; in their stead are strange things called Vertebrarias—angular, jointed, leafless stems, looking like small basaltic columns or the backbone of an animal. Judging from the print Dr. Falconer showed me of them, I should never have taken them for plants, but he says they certainly are such, though nothing is yet known by which they can be referred even to their class in that kingdom. With these are found many kinds of Ferns, a plant called Trizygia, supposed to be allied to Marsilea; leaves apparently of Palms, and other leaves resembling those of Bananas or Scitamineæ.

The book called "Vestiges of Creation," which is 1845. much talked of at present, has been attributed to several different persons, among others to Sir Richard Vyvyan, to my brother Edward, and to Lady Lovelace. Edward is certainly not the author; I cannot answer for either of the others. The account I have heard of the book is, that every really scientific man who has examined it, finds it shallow and unsound in his own particular department.

February 21st.

Anniversary meeting of the Geological Society;— Mr. Horner elected President; Wollaston medal awarded to Professor John Phillips (who is nephew of the famous William Smith); and Wollaston Fund donation to a Mr. Bain, an engineer, who has discovered the remains of an extraordinary genus of Saurians on the eastern frontier of the Cape colony.

Dinner of the G. S. at the Crown and Anchor; many distinguished men present.

February 22nd.

Went with Mr. Horner to one of Lord Northampton's great evening parties; met many of my friends:—Colonel Fox, Dr. Peacock, Mr. Milman,

1845. Sir Alexander Johnstone, Monckton Milnes, Sir
Charles Lemon, Babbage, and others.   It was
entirely a masculine party.   The Duke of Cambridge
was there; and a great *lion*, Mohun Lall, a Cash-
meree, who was poor Burne's interpreter at Cabool,
saved his papers, and was very useful to the English
prisoners.   On the table in the middle of the
principal room was a beautiful copy of the unfortunate Portland vase, one of those executed
long ago by Wedgewood.

February 23rd.

Spent the greatest part of the day with Fanny
and Leonora, at Mr. Stokes's (in Verulam Buildings,
Gray's Inn), looking over his splendid collection
of fossil woods.   It is extremely interesting.   He
is almost a universal collector, and his knowledge
is as multifarious as his collections, and seemingly
as deep and solid as it is varied.   His rooms exhibit
a most picturesque confusion of learned wealth,
literary, scientific, and artistical,—books, portfolios,
fossils, dried plants, stuffed birds, animals preserved
in spirits, pictures, busts, casts, coins, grotesque
figures from India or Japan, snuff-boxes, and nearly
everything that can be conceived ; and strange
to say, he seems generally able to find amidst this
chaos anything that he wants.

He is an enthusiastic admirer of Turner, and has
collected a number of fine prints to illustrate the
panegyrical book on that artist, written by Mr.

Ruskin. He shewed us five water-colour drawings 1845. by Turner, executed many years ago, truly beautiful, and quite free from that wild extravagance of colour in which he has indulged of late years. Altogether this was a most agreeable and interesting day. Mr. Stokes was in the highest degree, obliging, kind and agreeable.

Edward dined with us—much pleasant talk—discussed Puseyism, and the proceedings at Oxford against Mr. Ward;—Sir James Graham ;—and the debate in the House of Commons on the letter-opening ;—Turner's painting, and many other subjects.

February 24th.

Sydney Smith died the night before last—a great loss to the world, and especially to the world of London. He was certainly not intended by nature for a clergyman, but as a politician and a writer, his merits were very great.

February 26th.

Meeting of the Geological Society. Two good papers by Lyell on the Eocene and Miocene tertiary deposits of the coasts of the United States, followed by an interesting discussion. The geological formations of the United States are on a grand scale; the tertiary deposits extend without interruption down the coast from Maryland to Florida,

1845. and, as Mr. Featherstonhaugh afterwards stated, round the great promontory of East Florida and along the coast of the Gulf of Mexico, even to New Orleans, a length altogether, of not less than 2,500 miles, with a breadth in some parts of 150 miles.

The principal constituent of the Eocene deposit is a white limestone, which has often been supposed to belong to the cretaceous period, and which contains flints often very similar to those of the chalk ; but its fossils are decidedly tertiary. Sir Henry de la Beche compared this white limestone with that which occurs very extensively in Jamaica and other W. Indian islands, a considerable part of the latter he considers to be of the same period, though it occurs often under very different circumstances, the white limestone of the U. S. being undisturbed and nearly horizontal in position, and forming low plains while that of the W. Indies is often very much tilted up, and rises in some places to the height of 4,000 feet above the sea. Occasionally, he said, the Eocene limestone of Jamaica is covered by another white limestone lithologically undistinguishable from it, but of very recent formation, full of the same corals which at present compose the reefs in the neighbouring seas. Mr. Featherstonhaugh spoke of the extension of these tertiary beds to the S. and W. of the parts visited by Lyell, and mentioned that in East Florida, the white limestone which is found beneath the sand that covers the whole surface of that country, is, when first penetrated, as soft as soap, and may be cut by the spade into any shape, but by exposure to the air, becomes as hard as statuary marble.

A charming breakfast at Mr. Rogers's, the only guests besides ourselves were Mrs. Horner, Susan and Leonora. Rogers delightfully agreeable, full of good anecdotes and just remarks, of good sense, kindness and feeling. One wonders how he could ever have got the reputation of a severe, much more that of an ill-natured man; probbably, as Mary says, he has been much softened by age. He showed us some of the treasures of his very beautiful collection of works of art: some fine wood engravings from designs of Titian, very rough, but most spirited and masterly; original drawings of some of the greatest masters; a number of miniatures or "illuminations" of most exquisite beauty, cut out by the French from manuscripts in the vatican; and a collection of beautiful illuminations by the best Venetian Artists, prefixed to manuscript books of instructions, one of which was delivered by the Doge to every nobleman who was sent to take the command of a province.

Mr. Rogers's house is a perfect model of good taste. No ostentatious splendour, no overcrowding, no mere nicknacks or fantastic trifles, but a profusion of the most exquisite works of art, selected with admirable judgment. His powers, both of mind and body are in surprising preservation for a man considerably past eighty.

Visited the British institution for the second time this year. Some good pictures: and Irish merry-

1845. making (called "The Widow's Benefit Night") full of fun and spirit and character. "The Soldier's Dream," by the same artist, a very clever picture. Two large pictures by Martin—"Morning and Evening in Paradise,"—wild and exaggerated as is apt to be the case with that artist's paintings, but showing a grand and luxuriant imagination, with his usual remarkable power of impressing the idea of vast space. Some small paintings of dogs and game, by Landseer, not interesting in subject, but exquisitely painted. Beautiful sea-coast views by Cooke. A pretty little picture of Etty's—a Cupid floating in a shell on the water, and watching gold-fish.

March 2nd.

Mr. Brown* and Mr. Forbes† came to tea; the latter showed us a large collection of very interesting sketches of scenery made by himself in Lycia and the adjoining regions.

March 3rd.

A very pleasant dinner party at the Lyell's; the Edward Romily, Mr. Brown, Mr. Moore Esmeade. Much good talk on a variety of subjects:— the strange letter of Mr. Ward the Puseyite (published in *The Times*) in which he endeavours to show that his own approaching marriage is not inconsistent

---

* Robert Brown.   † Edward Forbes.

with the strong opinions he has published in favour 1845.
of celibacy ; the Bishop of Exeter ; mesmerism, etc.,
etc. Additional guests in the evening ; a very
agreeable party. Talked of fossil-botany and other
subjects of natural history, with Robert Brown, Mr.
Stokes and Mr. Broderip.

March 5th.

A very agreeable dinner party at Mr. Hallam's,
though he himself was unluckily called away by a
royal summons. Mr. Owen, Mr. Eastlake, Mr. Kay
Shuttleworth, Lord Landsdowne, Mr. Luttrell, Mr.
and Mrs. Horner, Katharine, and ourselves. I sat
next to Owen, and had much interesting talk with
him. Mr. Lowe and his observations on the natural
history of Madeira ; fishes of that island ; history of
fishes much neglected till of late years ; and those of
Guiana carefully studied by Mr. Schomburgh ;—
the largest fresh-water fish in the world, a native of
that country ; terrible fish in the rivers of S. America
which attack swimmers and tear off their flesh piece-
meal. Interesting book of Hugh Miller on the " Old
Red Sand Stone." Owen agrees with me in thinking it
the most fascinating book ever written on a
geological subject—contains the history of a powerful mind, as well as that of a geological epoch ;—
much to be lamented that such a man should have
sunk into the editor of a controversial newspaper,
and resigned science for polemical theology. Schomburgh about to publish a book on Guiana; valuable
work in the press by Count Strelezki, containing

1845. his researches in Australia. Owen thinks it will be the most important scientific work of this year; the Count's discoveries in the Palæontology of that country; bones of a species of Mastodòn found in caves, together with those of extinct Kangaroos and Wombats, and other Australian forms; rapidity with which the larger wild animals of Australia and Southern Africa are disappearing before civilized man; the wolf still holds its ground in France. Enjoyment of a naturalist in visiting distant countries, and finding himself among animal and vegetable forms quite unlike those of Europe. Dr. Andrew Smith—his enthusiasm for discovery; unfortunate that he has published no account of his expedition; exploration and civilization of Africa more likely to be effected from the side of the Cape than from the pestilential region of the Niger. A good deal of talk about the " Vestiges of Creation," which I perceive is now one of the most common topics of conversation. Owen agrees with the common opinion in believing Sir Richard Vyvyan to be the author, though he does not regard it as certain; Mr. Wheatstone thinks that it was principally Sir Richard's, but revised and modified by someone else; as, if it had been entirely by Sir R. Vyvyan, he thinks that it would have been much wilder. Owen said it was not very surprising that the author should have wished to conceal himself, as the theory which he has taken up is entirely opposed to the views which are generally held in England, as well by men of science as by the vulgar, and he might fear to bring odium on himself by his

opinions. The book has, however, been very successful. It is described as abounding with errors as
to the facts of science, but well written and
ingeniously reasoned, though on unsound data. Some
have fancied it was written by Brougham. That
eccentric genius is also said to have written a *novel*,
but none of the party could tell either the name or
the subject of it.

March 12th.

Called on Mr. Jones, the artist, who has apart-
ments at the Royal Academy. He showed me a
very good picture of the battle of Waterloo, which
he has been. painting for someone, and has just
finished—but he does not intend to exhibit it, from
a feeling which I think creditable to him, an unwillingness, while we are at peace and in professed
friendship with France, to irritate the French who
may visit this country—by ostentatiously displaying
our remembrance of an event so galling to their
national vanity.

Meeting of the Geological Society—a paper by
Sedgwick—comparison of the lower fossiliferous rocks
of North Wales and the Cambrian district. The
paper did not interest me particularly, but it was
followed by a very entertaining fight between Greenough and Sedgwick ; the former in his most caustic
style, attacked the paper as vague and unsatisfactory.
Sedgwick's reply was in his best manner, full of
rough humour and fun and strong good sense, and
pretty severe too ; but though there was plenty of

1845. hard hitting on both sides, there was perfect good
humour.

Heard of the death of my dear old friend, Miss
Fox, which grieved me much.   It cannot be
called unexpected as she has long been in a state
which deprived her of the power of conversing with
her friends, or even seeing any but her nearest
relations, and in which life could be no enjoyment to
her.

I can never forget her kindness to me, nor the
many cheerful, pleasant and instructive hours I have
formerly spent at her house, particularly in the
winter of 1835-6, when I was alone in London.

It is remarkable that Miss Fox and her two old
friends Sydney Smith and his brother, should have
died within so short a time of one another, all since
the 20th of last month.

Dined with Charles and Mary, and had much
pleasant chat.   I was much amused by an anecdote
Lyell told us, and which he had heard from old
Murray, the bookseller : Buckland's Bridgewater
(be it observed) is in fact a work on fossil remains,
and a very good one, but at the end of each chapter
there is a scrap of Theology dragged in awkwardly
enough, and tacked on like the moral to a fable ;—
well, a friend of Murray's had in his own copy
enclosed each of these Theological paragraphs within
brackets, in pencil, and had written in the margin
" As per contract."

I went afterwards with Lyell to Lord Northamp- 1845.
ton's and enjoyed it much.   There was a very large
party, and I met many I knew—in particular, my
old friend Stafford O'Brien, whom I was exceedingly
glad to see.   Also Sir William Hooker, who told me
that this extraordinary long and severe winter has
done much mischief to the shrubs (especially evergreens) in the open ground at Kew.

Dr. Falconer introduced me to Dr. Royle, the
great Indian botanist, author of an admirable work
on the natural history of the Himalaya mountains.
He is a large, fat, rather coarse looking man, not
prepossessing in appearance, but I found his conversation interesting.   He has long been engaged in
illustrating The Materia Medica of India, and is now
busy with the subject of Biblical Botany, in which
his extensive and intimate acquaintance both with
the productions and the languages of the East, and
his knowledge of the Arabian herbalists, and medical
writers assist him very materially.   He has satisfied
himself that the mustard tree of the New Testament
is Salvadora Persica, and the "hyssop," Capparis
Spinosa.

Among the curiosities on the tables at Lord
Northampton's, were specimens of common woods
(such as beech, larch, and Scotch firs), dyed of
various bright colours by means of metallic solutions,
absorbed into the tissue of the trees while growing.
I remember hearing something of this process in
1841, at the Meeting of the British Association at
Plymouth.   The colours are chiefly blue, green, and
various rich shades of brown.

March 16th.

Snow again.   It seems as if the winter never would
come to an end.   Yet in Cornwall there has been
no frost this whole winter, as Sir Charles Lemon
mentioned a few days ago.

March 19th.

We went to see the new Roman Catholic Church,
opposite Bedlam, built by Pugin.   It is very handsome, one of the finest modern specimens of Gothic
architecture that I have seen.   The carved work
particularly beautiful, the material, the limestone of
Caen, which is said to have been the stone most
used for ecclesiastical buildings in this country in
the Norman reigns.   A very rich window with
coloured glass at one end.   The Church is not quite
finished, but is expected to be consecrated in June.

March 22nd.

A pleasant dinner with the Mallets at Hampstead
—Mr. Prevost, Edward Romilly, Katharine and
Joanna Horner and ourselves.

Much talk about railways (which Mrs. Mallet
detests), Miss Martineau and Mesmerism, and the
recent detection of the imposture of the noted
clairvoyante, Jane Arrowsmith—the disturbances in
Switzerland—the Jesuits, and about murders, particularly the recent murders at Hampstead and Salt
Hill, and various points concerning judicial evidence.
Romilly and Mr. Prevost advocated the French

system of interrogating prisoners on trial. I opposed 1845.
it, as dangerous and cruel to the innocent. Romilly
mentioned a very curious case in which a man was
convicted of murder on circumstantial evidence,
which the judge who tried the case thought so
doubtful, that he wrote for a reprieve ; but before
the reprieve arrived, the prisoner confessed that he
had committed the murder, and detailed all the
circumstances of it—and every important particular
of the evidence on which he had been convicted
proved to be false. The result was right, and yet
the evidence which had led to it was all wrong.
The reprieve was stopped and the man was hanged.

March 24th.

The first day of fine weather.

March 27th.

I went to a sale at Stevens', in King Street,
Covent Garden, of objects of natural history ; bought
a set of Schomburgh's Guiana plants, about 500
species for £2 17s. 6d., the price to the original
subscribers being £2 10s per hundred species. There
were quantities of beautiful stuffed birds, and some
skins of antelopes and other animals. A skin of a
large boa-constrictor was sold for 8s. Two fine
antelope skins with the skulls for 11s., and two others
for 14s.

We went to the print room of the British Museum
and spent some time looking over the fine collection of engravings by Marc Antonio.

An evening party at Bedford Place, Mr. Brown
Mr. Stokes, Mr. Fellows, Mr. Babbage, Mr. Edward
Forbes, and many others.

I had a good deal of talk with Mr. Fellows, who
explained to me the reason of the great sickness that
occurred among the ship's crew employed in remov-
ing the Xanthian marbles—a sickness owing more to
imprudence and mismanagement than to any
peculiar unhealthiness in the climate of Lycia. The
lands between Xanthus and the sea are low and very
fertile, and after the first annual crop has been
taken off the ground, which is done by the beginning of May, the country people sow their grain
again among the stubble, turn the waters of the
river over it, and then return for a time to the
mountains with their flocks and herds ; the land thus
flooded bears a very rich second crop without any
further care, and the people come down again from
the mountains to cut it. Now the crew of the ship in
question, landed after the first crop had been
gathered in, and the water turned on. They pitched
their tents on the wet land, slept on shore instead
of returning to their ship at night ; indulged also in
considerable excesses,—and it was not wonderful
that fever should have ensued. Mr. Fellows said
that he had travelled repeatedly, and a great part on
foot, through Switzerland, Italy and Greece, as well
as our own islands, but the most beautiful scenery

he had ever beheld was that of the Southern parts of 1845. Asia Minor; the mountains are little inferior in grandeur to the Alps themselves, while the valleys and low plains have a richness and luxuriance scarcely equalled in the finest parts of Italy. In one day, he says, you may ascend from the region of the palm, and the sugar cane, to perpetual snow.

April 1st.

Went with Fanny and Susan to Mr. Eddis' to see the picture which he is preparing for this year's exhibition — the Infant Moses deposited by his mother in the cradle among the bulrushes. It is a truly beautiful and noble picture—almost the finest modern picture of a scriptural subject that I have ever seen—full of dignity, simplicity and feeling.

April 2nd.

Meeting of the Geological Society. The proceedings began with a short paper by Mr. Austen, giving an account of a supposed meteoric stone which had been found in a ploughed field near Lymington, after a thunderstorm; it was however extremely different from any aerolite hitherto recorded, being a mere piece of slaty micaceous sandstone, without any of the characteristic peculiarities that have been observed in every well-ascertained aerolite; and the evidence as to its origin seemed extremely weak. Warburton and Buckland attacked with great force the

1845. notion of its meteoric origin, and the general opinion of the Society seemed to be with them ; Buckland mentioned that he had met with two different persons who had actually seen the fall of meteoric stones, and in both cases the phenomena described were the same—a strong light moving with great rapidity, and becoming brighter as it approached to the moment of the fall, attended with a whizzing noise, but no explosion or sound like thunder.

The principal business of the evening was an important paper by Captain Bayfield, on the junction of the granitic and transition (Silurian) rocks in British North America. He traced this junction in a satisfactory manner along the whole chain of the great lakes, and the whole course of the St. Lawrence, from the Western extremity of Lake Superior to Labrador and Newfoundland, a distance of 2000 miles. There seems to be, as Lyell and Murchison noticed in their remarks on this paper, a most striking analogy between the geology of this vast tract of country, and that of Scandinavia and the northern parts of Russia, nay, many of the same fossil species which characterize particular beds of the Silurian system in England and Wales, are found to be equally characteristic of them on the shores of the great North American lakes.

April 5th.

Evening party at Lord Northampton's ; a great crowd. Sir Henry Pottinger and Mohun Lall were the chief *lions.* There was exhibited the new and 1845. very extraordinary process of "anastatic" printing by which an almost unlimited number of impressions of any engraving or printed page may be obtained with very little trouble or expense, and (it is said) without destroying the original.

In talking of copyright, to which this invention naturally led, Edward mentioned, that of the first edition of Thier's new book (Histoire du Consulat et de l'Empire), 10,000 copies were printed and were sold in one day; and 6,000 of the second edition were bespoken before it came out; and this in spite of a very large pirated edition which was published at Brussels almost at the same time. Dickens, it is said, made £2000 by his "Christmas Carol" which he wrote in three weeks.

April 16th.

Meeting of the Geological Society. A paper by Mr. Mackintosh, in opposition to the Glacial Theory as applied to North Wales, followed by a very good debate, in which Buckland, Sedgwick, Lyell, De la Beche, and Murchison, figured to advantage. After the discussion, Mr. Morris showed me some plates of fossil plants of the coal measures of New South Wales, which he has prepared for the illustration of Count Strelezki's book. Those Australian coal measures as he told me, like those of Burdwan, in India, are remarkably distinguished from ours by the absence of Sigillariæ, Stigmariæ and Lepidodendra.

1845.   The Glossopteris Browniana, a characteristic
fossil Fern of both the Indian and Australian coal-
fields, is so closely allied to the Glossopteris
Phillipsii of our Yorkshire oolite, that Mr. Morris
says he can find no satisfactory difference between
them. Lindley and Hutton discovered that
Glossopteris Phillipsii, when perfect, has a *digitate*
frond, (a very singular peculiarity), and Dr. Falconer
has ascertained the same to be the case in
Glossopteris Browniana. Mr. Morris is inclined to
think that the coal-fields of Burdwan and of New
South Wales may be of the Oolitic period.

<hr />

April 17th.

We made an excursion to the Botanic Garden at
Kew, a very pleasant one, as the weather had at last
become really fine. Vegetation however in the open
air is extremely backward, and there was little to be
seen at Kew except in the houses. These are in
great splendour, and I found innumerable objects of
interest. The house which contains the Australian
Proteaceæ and Leguminosæ is particularly rich :
magnificent Banksias and Dryandras of very many
species of great size, in the most luxuriant state of
health and vigour, and most of them in abundant
blossom ; Banksia speciosa in particular, quite a
large tree, bearing plenty of its singular fruits, other
species with their grand cones of yellow and tawny-
red flowers. Grevillea Baueri in blossom ; Grevillea
acanthifolia and Telopea speciosissima in forward

bud. Very few of the Proteaceæ of the Cape. 1845. Among the Australian Leguminosæ, the Kennedias and Hardenbergias were the most attractive. In another house, a great variety of New Zealand plants, and in particular the curious Phyllocladus, of which the branches seem, at one period of their growth, to be the midribs of pinnated leaves, bearing leaflets which strongly resemble the pinnules of a Fern. In another a splendid collection of Begonias, chiefly Brazilian, some of the beautiful Melastomaceæ of Brazil, the rare and curious Lace-bark tree, Lagetta lintearia, very lately introduced from Jamaica, the Upas from Java, the Maté tree of Paraguay, and the Poivrea coccinea (formerly a Combretum), a fine climber from Madagascar, with dense one-sided spikes of splendid scarlet flowers. The collection of Ferns and Orchidaceæ are also very fine ; among the former I particularly noticed magnificent specimens of Hemitelia horrida, two or three beautiful Lygodia, and the delicate Darea vivipara. In the Cactus house, the greatest curiosity was the Echinocactus Visnaga, an enormous misshapen monster of a plant, weighing above 700 pounds, and which was carried in a bullock waggon for some hundreds of miles over the rugged country of Mexico, yet reached this country in the healthiest condition possible. In the Palm house, I particularly observed two stately plants of a species of Agave or Fourcroya, with their immense pannicles of bulbs instead of flowers ; Caryota urens, Wallichia caryotoides, and Licuala—all remarkable for the singular shape of their leaves, and some of those

1844. gigantic succulent Euphorbias to which my eyes
became so well accustomed in the Cape Colony.

I had great pleasure in going through these
interesting gardens with my dear wife, who is well
able to appreciate them. It was a year and seven
days since I last visited them in company with her,
the day before we plighted our faith to each other.

_____

April 20th.

Charles Lyell, Katharine and I went to the
Zoological Gardens. Delightful weather, and the
animals seemed to be fully enjoying it. The greatest
curiosity here at present is the Sloth, which appears
to thrive very well. His attitudes and movements
are most singular, and unlike those of any other
quadruped ; he delights particularly to remain with
his back downwards, clinging by his hind feet to a
bar.

There is a white-headed eagle sitting on two eggs
—a very curious circumstance, as it is said to be the
first time that eagles have ever been known to breed in
captivity. She has made no nest, but merely scraped
together some straw and rubbish ; she looks extremely
fierce and vigilant, and is said to sit very steadily.
There is also a very pretty kid from Scinde, snow
white with bright silky hair, and a very engaging
countenance, and excessively long pendulous ears.

Crowned Pigeons from New Guinea, splendid
birds ; and a very beautiful Indian Antelope, the
common Antelope of the North of India, Dr.
Falconer says. The Giraffes appear to be in a very
healthy and thriving condition.

I visited the Exhibition of the New Society of Water-colour Painters, where there are several pretty things, but nothing that struck me particularly except the "Prisoner of Gisors," by Wehnert, the subject of which is a state prisoner intently occupied in carving figures on the wall of his dungeon.

Visited also the panorama of Nanking, which is very interesting ; the scenery beautiful and beautifully painted. The city, surrounded by its vast extent of walls, stands in the midst of a fertile and verdant plain, diversified with wooded knolls which remind me of those near Rio de Janeiro ; the great canal and the immense river Yang-tse-keang winding through the plain ; the view bounded by mountains, or at least steep hills of bold and varied forms. The famous Porcelain Tower is a conspicuous object in the foreground. The space enclosed within the city walls appears to be, like that within the walls of Rome, very densely peopled in one part, and almost desert in another.

Went with three of Fanny's sisters to see Sir Augustus Callcott's pictures, which are to be disposed of by private sale at his own house.

Called by appointment on Mr. Fellows, who
showed us his very interesting sketches and models
of the ancient buildings at Xanthus, which he discovered ; and afterwards took us to the British
Museum, to show us the sculptures brought from
that place, the arrangement and meaning of which
became very intelligible by the aid of his drawings
and restorations. The greatest part of the Xanthian
sculptures in the Museum belonged to a building
which Mr. Fellows believes to have been of a
triumphal nature, and erected to commemorate the
conquest of Xanthus by the Ionian Greeks, who
were mercenaries in the service of Harpagus, the
General of Cyrus. Of this building, Mr. Fellows
found only the basement standing ; all the super-
structure, the columns, statues and friezes had
been thrown down and scattered over the steep
face of the hill below, probably by an earthquake ;
and so completely were they hidden by thickets and
herbage, that when Mr. Fellows first visited the
place, the existence of only one of them was known
to the people of the country. The bas-reliefs, which
are now in the Museum, partly represent the
assault of a town (supposed to be Xanthus itself),
sallies of the besieged, battles, and captives brought
before a man in authority ; partly religious cere-
monies, hunting parties, and other subjects ; they
are quite Grecian in style, but appear to belong to
a ruder state of art than the Phigaleian marbles,
and Mr. Fellows, from various and strong considerations, supposes them to have been executed

nearly 100 years before the time of Phidias. The 1845.
beautiful, but unfortunately mutilated female figures,
in strong action and with flying drapery, belonged
to the same building, and appear to have been
placed alternately with columns. There appears
every reason to believe that all these were the works
of Ionian Greeks, the conquerors of Lycia; on
the other hand, the "Harpy Tomb," and the other
buildings of Xanthus, appear to have been the
works of the native Lycians, who in their own
language were called Tramalae.

April 25th.

Fanny, Joanna, and I went to see the New
Houses of Parliament, to which we had an order of
admission; then to see Thorwaldsen's statue of
Lord Byron, a noble work of art.

A very pleasant small party in the evening at the
Carrick Moore's.

April 26th.

We went to Babbage's evening party, which was a
very agreeable one, and less inconveniently crowded
than usual.

LETTERS.

To Lady Bunbury, his Stepmother.

2, Bedford Place,
April 26th, 1845.

My Dear Emily,

Maynooth is indeed *the* question of the
time, and I rejoice to see that there is little doubt of
the measure being carried, in spite of the extraordinary storm of bigotry that has been conjured up
against it. I certainly think it would have been a
much better plan if practicable, to unite the establishment for Roman Catholic education with that for
Protestant, and instead of endowing Maynooth, to
open Trinity College to all religious denominations ;
but I suppose that the measure proposed by the
Ministers was the best that was practicable *at present.*
In a few years, as Macaulay says, if Peel remains in
office, we shall have him proposing a regular
endowment for the Roman Catholic Church in
Ireland. Have you read Sydney Smith's posthumous
pamphlet on that subject ? I think it excellent with

the exception of two or three passages which are 1845.
rather too irreverent.—Of the speeches on this
Maynooth question, I do not think there was any I
liked better than Lord Howick's on Mr. Ward's
amendment. What a strange jumble and medley
was the minority on the second reading! made up of
bigots for the established church and bigots for the
voluntary principle, with some men (such as
Duncombe) whom one did not imagine to be bigots
for anything. I was much surprised to see the name
of Lord Charles Fitz-Roy in the minority. Perhaps
he voted in obedience to his dissenting constituents,
for it is said that a great majority of the petitions
against the measure proceeded from Dissenters.

Fanny is quite well and desires her best love to
you. We have been seeing a variety of sights—the
new Houses of Parliament the Panorama of Nanking,
Thorwaldsen's statue of Lord Byron, and above all
Kew Gardens; and have been lately at some very
pleasant parties, particularly at Mrs. Greig's and at
Mrs. Carrick Moore's. We are going this evening to
Mr. Babbage's. Our departure for Mildenhall is
fixed for the 6th of May. Edward is in very good
spirits, and busy with his classical articles. Lyell
has not yet got through his book.

Fanny has prepared a formidable artillery of
school-books and maps to batter the brains of the
children at Mildenhall.

I have seen my uncle William* once since he came
to town, but only for a few minutes as he was
excessively busy. I thought him looking very

* Sir William Napier.

1845. haggard and care-worn—I am very sorry for Henry's accident, but glad that my father is so well as you describe him.

Pray give my love to him.

Ever yours affectionately.
C. J. F. BUNBURY.

JOURNAL.

April 30th.

Meeting of the Geological Society : Murchison read a paper on the Silurian and older rocks of Scandinavia.

May 1st.

We went with the Lyells to the diorama, where we saw the views of Notre Dame at Paris, and Heidelberg, both very pretty. Afterwards Fanny and I went on to the Zoological gardens. We were sorry to learn that the Echidna which had arrived just before from Australia (the first of its kind that ever been brought alive to Europe) had died that morning.

May 6th.

Returned to Mildenhall.

LETTERS.

Mildenhall,
June 1st, 1845.

1845. My Dear Mr. Horner,

You will have heard of my excursion to Professor Henslow's, where I spent two days chiefly in looking over and selecting specimens of fossil plants, and brought away plenty of fine things to examine and describe and draw. The excursion answered very well, and would have been very pleasant in every way if the Professor had not barbarously thought fit to exclude my Wife from the invitation; we have however the satisfaction of knowing that he afterwards repented of his misdeed.

I shall have very great pleasure in seeing you here, at a time which, I rejoice to think, is now not far distant, and I am very happy that Mrs. Horner has resolved to accompany you hither. You will see Mildenhall in its best looks, for the season is so backward that the fresh and varied

green of spring still remains in all its beauty, and 1845.
the place looks more cheerful than at any other
season, and is besides enlivened by multitudes of
singing birds.

Our laburnums are just come into blossom, and
will not, I hope, be past by the time you arrive·
We have not many flowers, but Fanny is skilful
in making the most of those we have, and our
drawing-room looks gay and smells sweet with them.
We have at this season some very beautiful wild
flowers, but I am afraid the places in which they
grow are rather too far distant for you to walk to.
Since my return from Hitcham, I have tried my
hand at drawing fossils, and I think I succeed well,
but as this kind of drawing is new to me, it takes me
a long time; I expect I shall get on faster when
I have had more practise.

Is it not curious to see the *Times*, as it were,
patting O'Connell on the back in his opposition
to the government plan of colleges in Ireland, and
the bigots of both churches playing into one
another's hands against the cause of education?

The scheme proposed by Sir James Graham is
certainly conceived in a liberal and excellent spirit—
too liberal, I am afraid, for the present temper
of Ireland; but I think there are some objections
to the details of it. The greatest fault of the plan,
to my mind, is the making all the authorities and
professors of the colleges so dependent on the
government, thereby throwing into the hands of the
latter (if the scheme should be successful), a great
additional amount of patronage and influence, and

F

1845. tending to engender a subservient spirit in the seats of learning.

Again, when so moderate a sum (comparatively speaking), is available for the endowment of these institutions, is it advisable to divide that modicum between three different and distant colleges? Do you remember Lyell's objections against the number and poverty of the Scotch Universities? And would not his objections apply with equal force to the proposed Irish colleges? But it is perhaps useless to discuss objections to the plan, for as O'Connell and the Popish Bishops have thought fit to set their faces against it, I am afraid that though it may pass triumphantly through parliament, its operation will be practically defeated. There is one passage in the remonstrance of the Popish Bishops so absurd that it sounds like irony, but I am afraid it is serious bigotry; they say that the Roman Catholic students could not attend lectures on geology, anatomy, &c., without imminent danger to their faith and morals, unless priests of their own religion be appointed to the chairs of those sciences! This is fully a match for Mr. Sewell and his Christian morals.

I was neither surprised nor disappointed at the result of Lord J. Russell's motion; for I can see no good likely to ensue from a string of vague and general propositions, involving nothing practical.

Fanny and I unite in the warmest affection to you and Mrs. Horner and our dear sisters.

Ever yours,

C. J. F. BUNBURY.

To Charles Lyell at Kinnordy.

Mildenhall,
July 6th, 1845.

My Dear Lyell,

On our return hither from Barton, on Thursday, I found your book awaiting me, and I thank you most heartily for it. It is in every way a valuable present, and not least so as being the gift of one of the dearest and most valued friends I have on this earth. Yet I am almost sorry you have been so generous, for I fully intended to buy the book, and most certainly should not have grudged the money ; on the contrary, I should have been delighted to contribute my little share towards the sale of your work. I heartily hope it will have an extensive circulation. As I had already gone through the whole of the first volume (though I mean to read it again), I began the second, which most excited my curiosity, and I am very much pleased with what I have read of it. Your accounts of the coal formation of Frostburg and of the Ohio, of the Silurian rocks, and of the Lake ridges, are very interesting and satisfactory. If I had seen the Frostburg chapter before its final state indeed I might have suggested two or three corrections, but not of any great importance.

I have very seldom been more sorry to part from any one than I was from you and Mary at Cambridge. It was a delightful week, that of the Association,

F 2

1845. especially the last three days, and only passed
too quickly ; but I trust we shall have many more
such pleasant meetings. I can hardly reconcile
myself to the thought of not seeing you here at
all this year.

I think we must contrive by the help of the
rail-road, to run up to London for a day or two
before you go to America.

We enjoyed our visit to Ely (though there was
too much music for my taste), seeing its glorious
cathedral to the greatest advantage under the
guidance of our friend the Dean*; and at Barton
it was a great pleasure to find the *triad*† there in
unusually good health and spirits. I leave the
details to Fanny, who has such a talent for writing
them, and I know will not leave Mary uninformed
of any particular ; my way of epistolary writing is
more *sketching*, and it is lucky that our styles are
so different.

Sir Charles Lemon applied to me the other day
for the names of the best works on Fossil Botany,
and for some general information about the present
state of that science for the benefit of the Geological
Society of Cornwall. I sent him in return such
a general sketch as I could get into three sheets
of paper. Of course there could not be much
original in it, but I think that sort of thing is
just what I do best.

Pray remember me very kindly to Mr. and Mrs.

* Dr. Peacock.
† Sir Henry and Lady Bunbury and Cecilia Napier.

Lyell and your sisters. The recollection of the time 1845.
we spent at Kinnordy* is delightful to me.

Ever yours, most affectionately,

C. J. F. BUNBURY.

To Mr. Horner.

Mildenhall,
August 3rd, 1845.

My Dear Mr. Horner,

We are enjoying exceedingly, the visit of
the three dear girls, and I only wish it could last
much longer. I have taken many botanical walks
with Leonora, who is a very promising botanist, and
I have shown her most of the plants that are to be
found hereabouts at this season of the year.

Susan has painted a portrait of my Fanny, which
I think will please you as it pleases me extremely ;
it is, to my eyes, an excellent likeness, and a very
pleasing one ; the figure is very graceful, and the
composition altogether remarkably happy. She
painted also while we were at Barton, a very
clever sketch of a very old woman there, a favourite
of Cecilia's, who (the old woman, I mean) was much
delighted and amazed at her own likeness. Caricatures, as you may suppose, have not been wanting.

Cecilia had proposed to the sisterhood to toss me
in a blanket, but as they thought it (I suppose)

* On their wedding tour, June and July, 1844.

1845. rather too inhospitable to do so in reality, they
contented themselves with doing it on paper!

During our stay at Barton, I read Sedgwick's
article on the 'Vestiges,' but was rather disappointed.
I do not think it too severe, but too angry and
violent.

He is quite successful, I think in demolishing
the idea of transmutation of species, but that
fancy always appeared to me so utterly wild
and extravagant, that I did not need, for my
own satisfaction, to have it refuted.

I was much interested by your account of Malvern,
and can well imagine how happy you must be to
escape from business and cares, smoke and factory
questions, to the enjoyment of fine air and beautiful
scenery and geological rambles, beside the pleasure
of re-visiting a favourite haunt of former days. It is
twenty years since I was last at Malvern, but it is
very fresh to my memory, and there are few places I
should like more to see again. Some passages in
your letter re-called it most vividly to my mind.

I have a strong impression of its beauty, especially
of the views on the Herefordshire side, and I hope
that next year I may enjoy the pleasure of seeing
it in company with Fanny.

We are quite well, and I am happy to hear so
good an account of your health.

Believe me
Your affectionate Son-in-law,
C. J. F. BUNBURY.

To Leonard Horner, Esq.

Mildenhall,
October 22nd, 1845.

My Dear Mr. Horner,

I thank you very much for your kind letter of the 18th, from which I am happy to learn that you continue in good health. I can easily understand that your time is so mach engrossed by official business as to leave you no leisure for "epistolœ ad familiares;" and I heartily wish at once, for your own sake and that of your friends and of science, that you enjoyed more leisure. The proposed title of my paper is —— "Notes on some remarkable Fossil Ferns, collected by Mr. Lyell at Frostburg in North America." I fancy it is unnecessary to introduce into the title any reference to the speculations on climate, which occupy the latter part of the paper. I forget whether I told you that I have introduced into it my arguments in opposition to Lindley, in a more matured and digested form, for so few people heard what I said on the subject at Cambridge, that I thought it best to give my notions another chance.

My collection is as yet quite in its infancy, but is, I flatter myself, a thriving and promising baby, and kept very neat. Henslow particularly commended my care in labelling all my specimens, the want of which precaution, he says, he has remarked to be a source of great inconvenience in many private collections.

1845.   Henslow was with me when I received your letter, and I gave him your message. He and his eldest daughter came here the day before yesterday, and we got Mr. Hasted to meet them, but unluckily neither could stay more than one night. Their visit was however a great pleasure to Fanny and me, and we had much good talk. Henslow is still extremely occupied with the subject of Potatoes; he showed us the process of making *arrow-root* from them, which is a very pretty experiment and likely to be both attractive and useful to the poor. Mr. Hasted is very zealous for the establishment of a museum at Bury, and has some hopes of getting it effected; it certainly is rather a shame that such a town as Bury should be without anything of the sort. Both made many enquiries after you.

I think we shall come to London about the middle of next month.

We expect Edward to-morrow; he has been some days in the neighbourhood, staying at Sir T. Cullum's, and passing his time, I dare say, very gaily; in fact, he would have been with us yesterday, but for the attractions of a ball at Bury, which induced him to stay.

Cecilia has lent Fanny her pony, on which she took a ride to-day, I walking by her side. I have lately read Tristram Shandy to her in the evenings, of course omitting the improprieties; she was much pleased with Uncle Toby's character, but not, I think, so much tickled with the queer, quaint humours of the book as I am. We are now reading in the evenings, Coleridge's Wallenstein, which is new to me, but not to her, and with which I am 1845. delighted.

Mr. Basevi's death was indeed very sad, and must, as you observe, have been a great shock to the Dean. I recollect it was from Mr. Basevi that you got leave for us last spring to see the Conservative Club. Henslow told us that Lord Langdale was very near being killed in the same way, some time ago in the new buildings of the British Museum.

Pray give my best love to Mrs. Horner and Katharine. I look forward with great pleasure to our reunion in London.

Your affectionate son-in-law,
C. J. F. BUNBURY.

To Mrs. Charles Lyell.

Mildenhall,
October 30th.

My Dear Mary,

I write rather to remind you of my existence than because I have anything to say, for while you have been doing and seeing so many interesting things in America, our existence has been as uniform and unvaried as that of an oyster, and with almost as little incident. I was extremely interested by your letters which were forwarded to us to read, especially the last, in which Lyell gives such a curious account of the democratic institutions of

1845. Maine, and of their working. I confess I cannot
look on such a state of things as sound or good—at
least it would not be so in this country ; it can be
safe, I think, only where education is so widely
diffused as in the New England States, and where
there are such facilities and inducements for the
classes dangereuse to emigrate. Your account
reminds one a little of ancient Athens, but the New
Englanders are a much steadier people, and the
representative system, however democratized,
opposes checks to mere popular caprice and turbu-
lence, which were wanting in the democracies of
antiquity. What you say of the predominance of
asters and solidagos in the vegetation of New
England is exactly what I should have expected.
I am sorry I forgot to ask you to dry specimens for
me of the American oaks, and I suppose they will all
have lost their leaves by this time ; but if you should
meet with any, pray have the kindness to remember
this petition. Twigs with leaves will do very well,
though of course the acorns, if you should have the
opportunity of preserving them, will add to the
value of the specimens. The routine of our life here
is soon told. From breakfast till luncheon I
work at my botany and geology, and Fanny at her
old women ; after luncheon we go out walking, or
she rides on the pony Cecilia has lent her, and I
walk by her side. In this way she has made lately
some longer excursions than she was able to do
before, and yesterday in particular, we went to
Isleham, about five miles from hence, a most forlorn
and desolate-looking village, in a supremely hideous

and dreary country, but possessing a fine old Church 1845.
in which there are some very curious and well
preserved monumental brasses. Skye is our constant
companion in these excursions. After dinner we
each and severally take a nap ; then I read aloud to
her while she works. Yesterday evening we began
Tom Jones. When I am tired of reading aloud, I
betake myself to my geological studies, and Fanny
writes or otherwise employs herself till midnight,
which is our bedtime. For a week past her brains
have been so entirely full of *Coals*, that she could
hardly think of anything else, and as I tell her, her
head is a *carboniferous* formation, but there are no
fossils in it! Seriously however she has been
toiling most indefatigably and most meritoriously in
regulating the distribution of the coals which my
father gave to the poor ; you have no idea how hard
she has worked, and what pains she has taken to
find out the most deserving objects, and to distribute
the charity so as to do most good. I have been
delighted with the examination of some fossil wood
from Van Dieman's Land, which Lyell gave me in
the spring ; it is coniferous, and its structure most
beautiful, even under my microscope, which is a very
poor one. We had a letter lately from Henry, who
mentions having received one from you, and regrets
that he cannot go to Boston, but I suppose you will
also have heard from him, and therefore I need not
say more about his proceedings. This morning we
had a letter from my father. Edward was with us a
few days ago, in high health and spirits after his
German tour, and is coming back to us on the 4th

1845. of next month, for there is a " gunpowder plot "
against the pheasants which is to take effect on the
5th, and in which he is a party concerned.   He has
sundry antiquarian and historical irons in the fire,
and does not expect to go abroad before January.

God bless you, my dear Mary, and your excellent
husband, whom I love as my own brother; most
heartily do I pray that we may have the happiness
of seeing you return safe and sound to your own
country.

Your most affectionate brother,

C. J. F. BUNBURY.

P.S.—Fanny denies that she is unreasonably
engrossed by old women.   She has been planting
several new things in the garden.

JOURNAL.

November 21st.

We came up to London, and established ourselves in the Lyell's house in Hart Street, which
they have kindly lent us during their absence in
America.   Dined at Bedford Place, and met Robert
Brown, who is lately returned from the scientific
congress at Naples.   He seems to have been well
pleased with it.   There appears to be more scientific
activity, and more encouragement to science, in
that part of Italy, than I had supposed.

Went over to Bedford Place at luncheon time, and met Charles Darwin, whom I was very glad to see. I had scarcely seen him since June 1842, when we were together for some days in the Inn at Capel Cerrig, and made a great intimacy. He has long been an invalid, but has not the appearance of it, and is full of vivacity and ardour. We had a good deal of pleasant talk on scientific matters, especially on the geographical distribution of plants and animals.

He spoke of the extraordinary local peculiarity of the productions of the Galapagos islands, of which he has given a most interesting account in the new edition of his Journal; said, that nothing could be more striking than to see in all the plants and animals of these islands a well-marked South American aspect —a South American character as it were stamped on them all, while nearly all the species are peculiar. He avowed himself to some extent a believer in the transmutation of species, though not, he said, exactly according to the doctrine either of Lamarck or of the "Vestiges." But he admitted that all the leading botanists and zoologists, of this country at least, are on the other side.

Met at the Geological Society my old friend Major Charters, and had a long chat with him about

1845. old times and new occurrences. He is looking very
well and in good spirits, not at all altered since we
were at the Cape together, and very agreeable, as
usual. He seems to be rather rejoicing in the
probability of an approaching war with America, as
he says that it *must* come sooner or later, and that
we are just now in a good state of preparation
for it.

LETTERS.

To Charles Lyell, Esq.

16, Hart Street,
November 27th, 1845.

My Dear Lyell,

Here we are comfortably established in
your house, and very thankful to you for the use of
it; and everything I see about me reminds me
every minute of you and dear Mary, and of the
happy hours I have passed in your company. I
enjoy the change from the country very much, and I
hope to make good use of my time in London. I have
been extremely interested and delighted by your
letters, especially the account of the ascent of Mount
Washington, and its botanical characteristics. I
was not aware that Silene acaulis was a native of
America—it is not mentioned as such in any general

work on plants that I have met with. On the 1845.
Alps I have met with it at about the same elevation
as the top of Mount Washington, and about two
or three degrees farther North. You mention
also the avron or cloud-berry; this does not, I
think, extend nearly so far south in Europe;
neither does Diapensia Lapponica. I was aware
that the Azalea procumbens and the Linneæ
occurred in that part of North America, but the
extension of the latter to the southwards, till it meets
the Magnolia glauca, is a very curious fact. The
name of Mr. Oakes as a botanist, was not unknown to me. I shall be curious to learn the
result of his instruction about drying plants, as he
is so critical on those prepared by our botanists.
My paper on the Frostburg Ferns is to be read
next Wednesday. I have been unpacking two large
boxes of specimens from a coal mine near Oldham
which Mr. Horner got sent to me, and I find a
few very good things among them. I am struck
with the local peculiarities in the distributions of
fossil plants in the coal measures of this country;
each district seems to be characterized by certain
predominant forms, and the most common plants of
one coal field are rare or wanting in another
at no very great distance; for instance among
these specimens from Oldham, I find a great
prevalence of two or three species (or varieties) of
Neuropteris, and of several forms of Calamites;
Pecopteris muricata and abbreviata tolerably
abundant, but not a single morsel of Pecopteris
lonchites, which in general may be considered as

1845. the most common Fern of the English coal-fields, only one fragment of a Sigillaria and only one species of Lepidodendron (Lepselaginoides, I believe) but *that* tolerably abundant.   Stigmaria as usual.

Lady Holland is dead, leaving a will as strange and capricious as her conduct generally was in her life-time.   She has left (it is said) £1500 a year to Lord John Russell, who is to pay £500 a year out of it to the present Lord Holland.   £150 a year to her page.   £2000 (some say only £1000) to Charles Fox, it is said she has also left a sum of money (I have not heard how much) for a monument to herself. A good deal of sensation seems to be excited in the political world by Lord John Russell's letter to the citizens of London, in which he declares himself an advocate for the total repeal of the corn-laws; it is a very well-timed, manly, sensible and effective letter, and rather damaging to the ministers; the *Times* calls it a final death blow to the corn-laws.   One is curious to see what course Peel will take ; if he attempts to stick to protection, I think he will be thrown over. There is much anxiety too about the Oregon question, which looks very threatening, but I still hope that war may be averted.

I am very glad to hear that your lectures are so well attended, and that your book is liked by the Americans ; they must indeed be hard to please if they were not satisfied with it.

Now give my best love to dear Mary, and believe me,

Your most affectionate friend and brother-in-law,

C. J. F. BUNBURY.

Dined at Bedford Place—met a thorough
scientific party—Owen, Forbes, Dr. Fitton, and Sir
H. de la Beche, at dinner, Morris, Ansted, and my
brother Edward in the evening. Owen very lately
returned from the scientific congress at Naples,
h ghly delighted with his excursion. He was treated
with the honor he deserves, especially by the grand
Duke of Tuscany, who gave him a private interview
of half-an-hour, sitting by him and conversing with
as much ease and frankness, and absence of
constraint and formality, as any private gentleman ;
and when he took his departure, the grand Duke
made him a present of a most beautiful set of
models of the anatomy of the Torpedo, which had
been prepared for the Florence Museum.

The Museum at Pisa, Owen says, is the best
in Italy, at least in zoology. He spent a
week at Rome, in the palace of the Prince of
Canino, (Charles Buonaparte) who has a very fine
Museum of natural history, and is as zealous for
the science as ever. He possesses many interesting
relics of his family, derived from his grandmother
("Madame Mére") from his father Lucien, and his
uncle Joseph, who was also his wife's father. From
the latter he received a very great curiosity, the
original Charter of the Order of the Golden Fleece,
and the list of the original members. Owen quaintly
said—I suppose Joseph Buonaparte found it in his
carriage when he left Madrid !

At Naples there is but little encouragement for
science, and the few scientific men who are there,

1845. being generally poor, labour under great difficulties
in publishing anything, for no associations like our
learned Societies are permitted by that suspicious
government. The Neapolitans were quite amazed
that such an assemblage as the scientific congress
should be tolerated. The king however showed great
favour to that body, and paid some handsome compliments to the learned foreigners who were present.
Owen ascended Vesuvius, and I was much interested
by his account of its present state, a great alteration
having taken place since I was there in May, 1843.
At that time the crater had the form of an immense
pit, nearly circular, with extremely steep sides, and
at least 300 feet deep, at the bottom of which was a
comparatively small round of black slags, containing
the real mouth of the volcano, which continually
threw up jets of smoke (or steam) and red-hot stones.

At present, he says, the hollow is almost entirely
filled up, so that the top of the cone forms a rough
irregular plain, looking (as he expresses it) like a
petrified sea, and in the midst of this is the little
cone, now rising above the margin of the great one,
and still throwing out showers of scoriae. He was
struck, as I was, with the great similarity between
the noise issuing from the crater, and the puffing of
a steam engine, and he thinks that the phenomena
of volcanic eruptions are really occasioned by steam.
Vesuvius may at present be compared, he says, to a
steam engine with an overloaded safety-valve.

Owen is delightful—with such vast knowledge,
most unassuming simplicity, and a true unaffected
eloquence when excited by a favourite subject.

He is enthusiastically fond of the fine arts as well as 1845. of natural science, and spoke in raptures of the Italian pictures which he saw in his late tour, as well as of the scenery of the Alps.

Sir William Symonds called on us and talked very pleasantly. We had much talk about his fine museum of specimens of wood, &c. (mentioned in my journal of 14th of March last year), which he thinks of presenting to the British Museum. It is a collection unrivalled in its kind. He told us that the so-called "African Oak," long celebrated for its use in ship-building, has at last been ascertained to be a species of vitex; he had long sought to procure specimens of its leaves and flowers, and at last succeeded through the intervention of Mr. Soames, the great merchant and shipowner, who offered large rewards for them. The tree grows in the swampy forests far up the large rivers on the west coast of Africa, in situations inaccessible to Europeans on account of the pestilential air. He spoke also of the timber of a kind of acacia from Cuba, as most remarkable for its durability; a ship's keel made of this wood, after being in use for 87 years, was found, when the ship was broken up, to be in such perfectly sound condition, that it was taken as the keel of a new vessel. — Sir William Symonds laughs

1845. at Dr. Lindley's fancy that quick-growing oak timber is the best.

Meeting of the Geological Society : my paper on Fossil Ferns from Maryland was read, and was very well received. Sir H. de la Beche and Buckland, who were the principal speakers in the discussion that followed, expréssed great approbation of my paper, and gave me much encouragement. I was also much complimented afterwards by several others, and (what gave me especial pleasure), by Owen among others. But nothing gave me so much delight as my dear wife's joy and sympathy when she heard of my success. The discussion between De la Beche and Buckland turned on various points relative to the coal formations in general, and particularly on the *rolled pebbles* of coal, which in some localities (especially in South Wales), are found forming entire beds between the regular seams of coal.

Buckland conceived these to have been pieces of wood rolled into the form of pebbles (as we often see on modern beaches, at Cromer for instance), *before* they were carbonized ; while De la Beche maintained that they were actually rolled fragments of pre-existing beds of coal. Both agreed that they proved that the coal beds *above* which they occur, had been submerged and covered, either by the sea or by inundations before the foundation of the next solid seam of coal. Buckland remarked that something like a parallel to what we may suppose to have occurred during the carboniferous era, is now going on in Morecombe Bay (in Lancashire), where the sea has been gaining upon 1845. extensive peat mosses, sometimes covering extensive tracts of them.

Afterwards, Dr. Mantell read an interesting paper on the Wealden strata at Brook Point in the Isle of Wight, and the fossil remains found in them. Immense quantities of fossil trunks and branches of trees, often of great size, lie on the beach and imbedded in the cliffs, heaped together, and laid across one another without any order; their bark is carbonized, their wood is of the carboniferous structure, agreeing with that of the Araucarias. mineralized by carbonate of lime, and much incrusted and penetrated by pyrites. Among these are found occasionally cones, numerous fresh water shells, of a species which Mantell calls *Unio Valdenses*, and innumerable bones of huge reptiles, especially of the Iguanodon. Dr. Mantell justly observed that this remarkable accumulation of organic remains presented a most striking analogy to those great natural drafts of drifted wood, which are so frequent in the Mississippi and the great South American rivers, and which carry down with them the bodies of numerous quadrupeds and reptiles. He suggested also that the abundance of pyrites encrusting the remains, might owe its origin to the development of sulphurated hydrogen from the decaying animal matter.

Some of the bones of Iguanodon procured from this locality and exhibited on the table were of

1845. astonishing size, exceeding any that had been seen before.

Another great curiosity, collected by Dr. Mantell at Brook Point, was a specimen of that singular fossil plant, the *Clathraria Lyellia*, exhibited its internal structure, which had been hitherto unknown. It has been examined by Robert Brown, who pronounces it to be unlike any plant previously known, either in the recent or fossil state, but with some general and vague analogy to the Cycadeæ —certainly not endogenous.

After the meeting I had some talk with Major Shadwell Clerke, who does not believe that the Americans will go to war with us.

December 4th.

The *Times* this morning asserts as a positive fact that parliament will be called together early in January, that the Queen's speech will recommend an immediate consideration of the Corn Laws, with a view to their total repeal, and that Wellington, in the one House and Peel in the other, will without delay bring in measures to that effect. This is indeed important news, if true. I should like to see the faces of our Suffolk members and the rest of the farmers' friends at this announcement.

A very pleasant evening party at Bedford Place: Mr. Brown, Mr. Forbes, Mrs. Richard Napier*, the

* Emily, the Wife of Lady Bunbury's Brother, a very charming person.

Phillimores, Mr. Babbage, &c. Mr. Young, who is 1845.
a Tory, declares that his city friends entirely
disbelieve the statement in the *Times* about the
repeal of the Corn Laws ; but their wish is evidently
father to the thought. Babbage believes the *Times*
is right.

Forbes told me that beds of good coal, which was
expected to prove very productive, have been discovered in Borneo, and that the fossil shells found
in company with it, have proved to belong to the
Cretaceous series. The coal which occurs near
Santa Fé da Bogata, in New Granada, has likewise
been ascertained to be of the Cretaceous period. He
remarked that there appears no reason why deposits
of real and good coal should not have been formed
in every one of the geological eras, though all that
in our own country belongs to one period. The
coal of Northern India and of Australia appears
to be most probably of the Oolitic age. Agassiz,
it seems, is about to publish a book in which he
intends to prove that all the Tertiary fossils which
have been hitherto identified with recent species,
have been wrongly identified, and that in reality *no*
one shell is found both in a recent and fossil state.
Forbes thinks that he will only prove his own want
of practical acquaintance with the variations of
living shells. These local variations, he says, are
very apparent to any one who has much practice
in collecting shells in their native localities, so that
wherever you find a variation in the state of the sea,
the currents, &c., nay, as often as you pass a
marked promontory on the coast, you will find

1845. different varieties of the same species. In the common oyster these variations are so observable, that those who are much accustomed to them can tell at once, by the appearance of an oyster, from what part of the coast it is procured, and also what is its age.

December 5th.

The *Standard* indignantly and vehemently denies the truth of the assertion that Peel intends to repeal the Corn Laws ; the *Times* as positively repeats its assertion. It will certainly be droll, if the very House of Commons which was elected expressly to maintain the Corn Laws, should be made the one to repeal them.

December 6th

Went with Mr. and Mrs. Horner, Leonora, and Joanna, to see the Museum of Economic Geology, which is at present in Craig's Court. Sir H. de la Beche with great good-nature accompanied us through the whole of it, and gave up a great deal of his time to us; but he talked at such a rate and carried us from one subject to another with such rapidity, and through such a variety of subjects that I found it difficult to follow him, and impossible to retain much of what he told us. The collection is certainly very interesting ; one ought to make many visits to it, and take a particular department

of it each time. There is an ample series of 1845.
specimens of all the kinds of stone useful in building
and found in England, the label of each specimen
stating not only its locality and mineral nature, but
also the principal buildings in which it has been
employed; polished specimens of all the British
granites, serpentines, and marbles, some of which
are not inferior to the finest foreign kind (the red
granite of Peterhead, for example, is quite as beautiful as the Egyptian) ;— fine illustrative series
of the ancient and modern kinds of earthenware,
porcelain, glass, and enamel, with the substances
used in their manufacture ; different varieties of
bronze, and some beautiful English specimens of
casting in that material ; illustrations of the electro-
type process ; beautiful sets of electrotyped planis
by Mr. Ibbetson ; series illustrating the different
varieties of copper, iron, and steel, and the processes
employed in their preparation ; and among other
things in this department, specimens of that curious
change in the texture of iron (from fibrous to coarsely
crystalline), which takes place in iron axles under
particular circumstances, and which, by rendering the
iron very brittle, has been the cause of several
accidents on railroads.

Also many interesting geological models, on the
ingenious plan invented by Mr. Sopwith, models of
mines, and of all kinds of mining machinery, and
a variety of other interesting and curious matters.

De la Beche told us, that the architects of the
Norman period in this country, seem to have been
particularly careful and judicious in the selection of

1845. durable stone for their buildings, and that in the
succeeding ages, less and less attention appears
to have been paid to this point, the facility with
which stone could be carried, being more considered
than its durability.

December 11th.

Fanny went out to pay some visits (for I was
detained at home by a bad cold), and she brought
back the extraordinary news of the resignation of
Peel and his ministry. It seems that the Duke of
Wellington and other members of the Cabinet were
obstinate on the subject of the Corn-laws, and Peel
determined to resign rather than meet Parliament
without proposing the repeal of those obnoxious laws.
Lord John Russell has been sent for to form a new
administration. This is an unexpected and startling
change. A month ago, no one would have dreamed
that Peel's administration was not as firmly rooted
as any ministry can be. Yet it is perhaps a change
of individuals rather than of principles, for the Peel
Conservatives have of late become so very like
Whigs, that the difference between them seems little
more than nominal. I am afraid that Peel, now
he is out of power, may fall back a little towards
Toryism, and altogether I do not expect that the new
ministry (if formed of Whigs only) will last very
long.

Went to Sir William Symond's model room, at Somerset house, to see his collection of woods. The set from New Zealand is particularly rich. Among other things, I noticed several species of Podocarpus, the Totarra, ferruginea and spicata (figured in Hooker's Icones Plantarum), and the curious Phyllocladus trichomanoides, of which the wood is very hard and compact. In these trees and also in the Dammara or Kowdie tree, the annual rings of growth are very distinct, but very narrow. The wood of the Knightia excelsa (one of the Proteaceæ) is very beautiful ; the medullary rays large and wavy, as in the Proteaceæ of the Cape.

Here is a specimen of the Cedar from the mountains of Morocco, but Sir William believes it to be the Deodara, and not the Cedar of Lebanon, chiefly from the smell of the wood. The small twig with a few ill-preserved leaves which is in this collection, is not sufficient to determine the species. He has a box made of the wood, and he says that persons familiar with the Cedar of the Himalayas have recognized it at once by the smell. The Acacia from Cuba, which I mentioned the other day, and which has been found so very durable in shipbuilding, is called Sabicu. Another timber from Cuba, equally valuable, and which has been known for some time as the Cuba Cedar has been ascertained to be Cedrela odorata, which by the way, is also called Cedro in Brazil.

The Cork tree, though so very similar in its leaves and fruit to the Ilex, appears to be materially

1845. different in the texture of its wood, as well as its bark; its wood is loose-grained with very large medullary rays.

Sir William says that Lindley is quite mistaken in supposing the Sardinian Oak, which is used in ship-building, to be the same as the Quercus Cerris; the wood of the latter is quite useless for that purpose. Larch timber, he tells me, is one of the best for ship-building, more durable than English Oak, and nearly as much so as Teak. It is a curious fact that Oak when joined with other woods in the construction of a ship, appears to corrupt them; a ship built partly of Oak and partly of Teak, becomes useless much sooner than one built entirely of Oak.

December 14th.

Nothing known yet as to the formation of a new ministry. It is rumoured that Lord John has declined to undertake the administration.

December 17th.

Meeting of the Geological Society; four papers read, but short ones, and of no particular interest. The first was Owen, on some bones from the Wealden formation, which he had formerly referred to wading birds, but having since cleared them more thoroughly from the matrix, and examined them more minutely, he has come to the conclusion that they are more probably bones of a Pterodactylus than of a bird. Dr. Mantell (who was the discoverer of the said bones), made some observations.

The second paper of the evening was on an

interesting subject, but too brief to be satisfactory ; it 1845.
was an abstract communicated by Murchison of
Professor Göppert's researches on Amber, which,
according to Göppert, is the resin of an extinct
species of Fir, nearly allied to the Spruce Fir. The
extensive deposits of it which are known to exist on
the south of the Baltic, are said by the author to have
been drifted from forests now covered by that sea, and
their geological age is believed to be that of the *Molasse*.

Among the substances enclosed and preserved
in amber, Göppert has ascertained a great many
species of plants—not less than 48, if I remember
right—principally dicotyledonous, but some cryptogamous, and among the number some mosses, a
tribe of plants extremely rare in a fossil state. The
plants preserved in amber are mostly of existing
genera, but all of extinct species, and have more
analogy on the whole to the recent Flora of North
America than of Europe. Insects as is well known
are found in great number and variety in the amber
of the Baltic.

After this, was read an extract of a letter,
describing a late earthquake in Cutch, by which a
pretty extensive submergence of land had been
occasioned. Sir H. de la Beche made some
remarks on the small change of level which would
be sufficient, on some parts of our own coast, to
cause extensive submergence. Lastly, there was a
short communication from Buckland, on some round
bodies which are found on the shores of Lough
Neagh, and are called by the country people
" petrified potatoes." He exhibited some of them,

1845. and explained his idea of their origin, that they were masses of clay or marl, rolled into a round shape by the action of the water, and which in that process had picked up small pieces of stone of various kinds, which were now partly imbedded in them, and partly adhering to their surface; and that they had afterwards become indurated. This explanation seemed very probable, but De la Beche dissented from it, considering them as concretions, analogous to *Septaria*. Mr. Bowerbank mentioned a curious fact, that among the fossils of the London Clay at Sheppey, are found real tubers of some plant in a fossil state, much resembling those of Oxalis crenata, and showing, when slices of them are examined under the microscope, their cellular tissue in good preservation.

December 18th.

It is said that among other legacies, Lady Holland has left 300 *fans* to Lady Palmerston, all her dictionaries to Mr. Charles Howard, and a great collection of caricatures to Mr. William Cowper.

The negociations for the formation of a new Ministry are still going on, and nothing settled yet. I am told that one difficulty is the opposition of Lord Lansdowne and Lord Palmerston to the total repeal of the Corn-laws.

December 19th.

It is at last announced, that Lord John Russell has definitively accepted the charge of forming a Ministry.

LETTERS.

16, Hart Street,
December 31st 1845.

My Dear Lyell,

A happy new year to you and dear Mary, 1845.
and plenty more of such—though I hope your New
Year's Days will not always be spent at such a
distance from us—I had just finished writing to you
the beginning of this month, when we received
Mary's kind and pleasant letter, for which many
thanks to her; and we have seen several more of
her letters since, by which we hear that you had got
as far as Philadelphia on your southward journey.
It is delightful to have such uniformly good accounts
of you both, and to find that you are spending your
time happily in America. I am in hopes that the
prospect of a war between the two countries is now
much diminished, and everybody seems to agree
that even if war should ensue, it will not happen
this year, so it will not interfere with you. Stokes
says, he defies England and America to go to
war about the Oregon.

1845.   You will I suppose have seen enough of the
English newspapers to have some idea of the very
strange ministerial crisis that has occurred here :—
how Sir Robert Peel on a sudden resigned office, to
the astonishment of everybody ; how Lord John
Russell was sent for to form a ministry ; how, after
long hesitation and consultation he undertook it ;
how the attempt to form a Whig cabinet failed
through the insuperable objection of Lord Grey
against Lord Palmerston ; how Peel and his cabinet
have been re-instated with no other alteration (as
yet) than the departure of Lord Stanley (a good
riddance certainly), and the substitution of Mr.
Gladstone for him. Peel, as usual, keeps his real
intentions very secret, but almost everybody believes
that he does intend to propose the total repeal
of the Corn-laws, and that the refractory members
of the cabinet whose opposition induced him to
resign, have submitted to his ascendancy. It is
said that none of his colleagues except Sir James
Graham and Lord Aberdeen were with him at first
on this point. I received at the beginning of this
month a letter from Mr. Dawson of Picton in Nova
Scotia, accompanied by a very good paper, which
he wished me to bring before the Geological Society,
but the specimens which were to illustrate the
paper did not arrive till the 24th. They are not of
much value except in connection with the paper, but
in that there are several interesting observations,
especially on stigmaria and sternbergia.

He sent me also a few Nova Scotia mosses and
lichens, all of them I think European species.

My paper on your Frostberg Ferns was read 1845. before the Society on the 3rd, and was very well received—indeed I received a good many compliments on it. At the last meeting there were several short papers, but nothing of much interest.

Henslow wrote some time ago to Mr. Hutton of Newcastle, mentioning my design of a work on fossil botany, and suggesting that he should assist me. He returns for answer that he thinks of resuming the " Fossil Flora " himself, and that he thinks it impossible for anyone to carry on such a work with advantage unless he lives in the midst of a carboniferous district. So it is clear that I am to expect no assistance from him—but his rivalry would be formidable. I hear that Agassiz has written a book in which he *thinks* he has proved all the tertiary shells are specifically distinct from the recent, and that all the identifications hitherto made are erroneous. Forbes thinks he will only prove his own want of practical acquaintance with the variations of living shells. You will be very sorry to hear as I was, that poor Forbes is now very seriously ill—I hope not dangerously. He would be a grievous loss to science as well as to his friends.

When the alterations of the British Museum will be finished, I cannot imagine ; at present it is a most confused looking medley of new and old ; the old gateway is not down yet, but the entrance is shut up, and one goes in by a side way, and wanders through endless passages between temporary boardings and through galleries of half arranged objects till one is almost bewildered. Your enemies the architects

H

1845. seem to have had their will with a vengeance there—
I almost despair of ever seeing it in a satisfactory
state of completion and arrangement. I suppose
you have by this time got far away from the lands of
frost and snow, into the region of magnolias. Here
we have, as yet, a very mild and very wet winter,
singularly mild without any ice or snow hitherto—
perhaps they are to come. This agrees very well
with me, but not with many people, and I think
not quite with Fanny—I hope you are thoroughly
enjoying your tour, and will make large additions
to our knowledge, especially when you reach the
western states but above all I hope you will both
return safe and sound from your expedition. Best
love to dear Mary.

Ever your affectionate brother-in-law,
C. J. F. BUNBURY.

JOURNAL.

January 13th.

1846.   I do not intend this Journal for a register of public
events, which are sure to be sufficiently recorded in
printed documents   therefore I need not dwell on the
strange ministerial crisis (one of the strangest I
remember in my time) which ended in the re-estab-
lishment of the Peel Cabinet, with only one or two

changes in its personal composition. It seems to be 1846. certain that the cause which defeated Lord John Russell's attempt to form a ministry was the suddenly declared antipathy of Lord Grey to Lord Palmerston. One may well ask why Lord John could not have formed a Whig ministry without Lord Grey; but there is no great harm done; though it might have been better for the fame of the Whigs if they had given the death blow to the Corn-law, yet it is just as well for the nation that it should be done by Peel who is likely to do it as effectually and more easily.

The Protectionists, indeed, in the eastern and southern counties, are making a desperate noise, holding innumerable meetings and making endless and furious speeches, but are only exposing the weakness of their cause, of their want alike of popular strength, sound argument, and of oratorical power. They lay themselves sadly open to the keen satire and ridicule of *The Times* and of *Punch*, and do more to bring their order into discredit and contempt than the league could do. It seems clear that the doom of the Corn-law is fixed. But neither Peel nor the Whig leaders deserve the credit of the victory over them. It has been won by the middle classes headed by Cobden; for the working people, even in the manufacturing districts, appear not to have always co-operated with them.

But now even the agricultural labourers, as it seems, are beginning to open their eyes to their real interests in this matter. The most striking sign of the times has been a meeting of labourers in an obscure village in Wiltshire, where they expatiated on

1846. the misery of their condition, and showed how little they had been benefited by the Corn-law, and how little they had to fear from the repeal of it.

I still think that a moderate fixed duty on foreign corn would have been best, and think it a great pity that the opportunity for such a compromise was not seized when it might possibly have put an end to agitation on the subject, and conciliated the manufacturers without any great shock to the landowners. But now it is clear that the time is past for such a compromise, which would now *settle* nothing, and in the state to which affairs are brought, the thing most desirable, even for the agricultural interest, is *certainty*, a cessation from strife and agitation and doubts and fears.

To return to my own concerns, I have, by God's goodness, entered on a new year of my life, and I heartily pray that at the close of it, I may be a better and wiser man, and less weak than I am at the beginning. I have much cause to be thankful for the happiness enjoyed during the past year, which on the whole deserves to be called one of the brightest in my life.

I have very much to do in the improvement of my temper, and strengthening of my mind.

On the 4th of this month (immediately before the attack of influenza) I began to read the Life of Blanco White, which promises to be deeply interesting. Yesterday also I made a fresh start to begin the study of German, and I hope I may have more perseverance in that pursuit than I showed last year.

But the study of languages is strangely distasteful

to me. I ought to feel myself stimulated by the 1846.
example of Blanco White, who began to learn Greek
when more advanced in life than I am now, and
though he could spare only a quarter of an hour per
day for it, persevered till he had read through many
of the principal authors in that language.

Went to the Linnean Society, and spent some time
in studying the very valuable botanical disquisitions
of Mr. Brown and Mr. Bennett in the "Plantæ
Javanicæ Rariores." Then I looked through the
numbers of the " Annales des Sciences Naturelles,"
and the "Annals of Natural History" for the past year
to see whether they contained any information on
the subject of fossil botany, as I have promised Mr.
Horner to give him an account of what has been
done in that branch of science since 1844. I found
nothing on that subject, but I read with much
interest in the " Annales des Sciences Naturelles,"
Schouw's paper on the Geographical distribution of
the Coniferæ in Italy. Thence I went to Bedford
Place, chatted awhile with our sisters, then came
home, and read German with my dear wife till
dinner.

In the evening read several chapters of " Tom
Jones " to her, and afterwards went on with the
" Life of Blanco White."

1846.   January 16th.

At the Geological Society, whither I went to read,
I met Dr. Falconer, who told me that the City
people are very uneasy about the news arrived from
America by the last mail, and that it is expected to
have a serious effect on the funds. There is
certainly something very unpleasant, and alarming
to the friends of peace in the extravagant pretensions
put forward by the democratic party in America,
and the blustering and arrogant tone in which those
pretensions are expressed. It may be mere unmean-
ing bluster, or it may be meant merely to serve a
party purpose in their own internal politics, but
it may also lead to much more serious consequences
than the blusterers really contemplate.

January 17th.

A very pleasant party at Bedford Place · the
John Moores at Dinner, and in the evening, Miss
Moore, Mr. Stokes, Sir William Symonds, Professor
Phillips, Edward, etc. John Moore is a very great
favourite of mine.

I had some talk with Mr. Stokes about American
affairs, in which he is particularly interested. He
does not speak so confidently of the maintenance of
peace as he did a few weeks ago ; he still thinks that
it is not the intention of the American government
to engage in a war, but there is danger that such
feelings may be excited in both nations by their
declamations and boasts and invectives, as may

ultimately render a war unavoidable. He says that 1846
General Cass, whose violent and boastful harangue
against England has just now excited so much
sensation, is labouring to ingratiate himself with the
Western people, in hopes of being hereafter
raised to the Presidency by their influence; and
that he seized the opportunity of making that clap-
trap speech in the absence of Webster and Calhoun,
from either of whom if they had been present, he
would have received a severe castigation. I was
much surprised to hear from James Heywood, that
there is a strong belief among his Manchester friends
that Peel will after all propose a compromise (a fixed
duty on foreign corn, to last five or six years, and
then to cease altogether), and what is more, that
the manufacturing interest will not make any strong
or general opposition to it. Now, James Heywood
may be picking up and reporting the opinions of
others.

January 19th.

An evening party at the Richard Napier's.
Charades were acted by Augusta Napier, Susan
and Joanna Horner, Charley and Dick Napier, and
Edward; they were exceedingly well got up, and
acted with much spirit. Dick Napier, a very fine
spirited and clever boy, acted capitally. Among
the audience, besides ourselves and the party from
Bedford Place, were the John Phillimores, Mr. W.
Grey, Sir Thomas Bourchier, old Lady Charleville,

1846. etc. It was a very pleasant evening. I had some talk with Henry Napier about his History of Florence, a great work, which he has just completed, but which I fear he will find much difficulty in publishing, as he cannot afford to bring it out at his own expense, and no publisher is likely to be adventurous or liberal enough to undertake such a work. He told me that for nine years he had devoted from nine to fourteen hours a day to this employment, and that for the last volume alone (containing the reign and legislation of the Grand Duke Leopold), he had looked through 300 folio and quarto volumes on law, exclusive of all other subjects.

Mr. Horner told us he had heard from an M.P. whom he believed to possess good means of information, that the plan which Peel intends to propose in regard to the Corn-laws, is a duty of six shillings a quarter on foreign wheat to be reduced by 2s. each successive year, and thus to expire at the end of three years. Still I cannot but doubt whether the league, having carried on their warfare till they are so near to a complete victory, will at this time of day consent to any compromise, and even whether the agricultural party will think such a palliative worth accepting. But a very few days will now show the truth, as Parliament is to meet on the 22nd.

January 21st.

Meeting of the Geological Society. First was

read the conclusion of Sedgwick's paper on the 1846. Geology of Westmoreland, the first part of which I had missed, having been absent from one meeting of the Society on account of influenza. This concluding portion was rather unintelligible, and consequently uninteresting to me, depending on local details, and on an intimate knowledge of the characters and sub-divisions of the Silurian system.

Nor could I well follow, nor take much interest in the discussion that ensued, principally between Mr. Daniel Sharp and Professor Sedgwick and which turned mainly on the question whether certain beds belonged to the upper or lower Silurian group. Afterwards was read a paper by Mr. Dawson, on the stigmaria, and other fossils from Nova Scotia to which I had added some botanical notes, but there was no discussion on it.

Sir Henry de la Beche, whom I had expected to oppose the *root* theory of the stigmaria, was absent, and in fact it happened that there was no one present who had paid any attention to Botany, except Mr. Morris, and he was too diffident to speak. Dr. Mantell only spoke, and expressed with great positiveness his conviction that stigmaria was suffi-ciently proved to be the root of sigillaria.

After the reading of the papers, I had some talk with Mr. Pratt, a distinguished geologist, who promises to show me a collection of fossil plants that he has procured from coal mines near Oviedo. These coal mines which an English company has undertaken to work, promise, he says, to be very rich ; they are worked with great ease and con-

1846. venience, by means of *adits*, at different levels in the sides of the hills, no shafts being required.

They are in a lower position, geologically speaking, than any of our English coal mines, probably even in the Devonian system; and what is curious, below the beds of solid coal occurs a vast bed of a conglomerate consisting of rolled pieces of coal. There occur also enormous masses (or beds, I am not sure which) of red iron ore, yielding 75 per cent. of iron; but the clay ironstone, so general in our coal fields, is not met with. The fossil plants, Mr. Pratt says, are not numerous, and belong chiefly to sigillaria, lepidodendron and calamites; he has found only one single specimen of a fern.

———

January 23rd.

Read the debate in the House of Commons on the Address in answer to the Queen's speech—interesting though the serious struggle has not yet begun, for there was no amendment. Peel made a full and fair recantation of his former opinions in favour of Protection, and avowed himself a complete convert to Free Trade, though it would appear that he does not mean to propose a total and immediate repeal of the Corn-law. Lord John Russell's explanation of the circumstances attending the abortive attempt to form a Whig ministry,—not very satisfactory. It is pretty clear, I think, that Lord Grey's resistance was only the *ostensible* reason, an excuse, in fact, for withdrawing from an enterprise 1846. which appeared hopeless.

A clever, biting, virulent speech of D'Israeli against Peel: he seems inclined to make himself the mouthpiece of all the bitterness and venom of the disappointed Conservatives and Protectionists, and his talents are sufficient to give him the power of wounding severely.

Peel promises to expound his scheme of Free Trade on Tuesday next, and then will begin the grand struggle, — the most important political struggle that has occurred in my time ; its practical consequences are likely to be more important than those of Catholic Emancipation, and far more so than those of the Reform Bill. The newspaper took up much of my time this morning, but I read about 40 pages of Edward's essay on the Topography of Rome, in the Classical Museum ; also some pages of Goppert on the structure of recent and fossil ferns (in German), at which I have been working for some days; in the evening read between 40 and 50 pages of Blanco White, partly aloud to my wife.

We are both much struck by the generous, gentlemanlike, and Christian spirit of Blanco White's reply to a bitter and ill-tempered letter of John Allen's, on his book against Popery (volume 1, page 416).

Read a good spell of Goppert, and made some notes from it. Finished the first volume of the the Life of Blanco White, a most interesting book, and one which makes a deep impression on me.

I do not think I did justice in my yesterday's journal to Feel's speech, which was very manly, clear, and vigorous, and brought forward very convincing arguments from experience in favour of Free Trade.

Read 70 pages of the second volume of Blanco White. Read prayers at Bedford Place with Mrs. Horner and our sisters. Saw there some numbers of the new newspaper (the Daily News,) which is partly conducted by Dickens : it promises very well. A very amusing account in it by Dickens of his journey in France.

Went on again with Goppert. Spent some time at the Linnean Society, and made several extracts from the "Annales des Sciences Naturelles." Finished, with my wife, the second act of " Die Deutsche Kleinstädter," and in the evening read to her about 30 pages of Blanco White.

Goppert again, read Herman Merivale's review, in the Edinburgh on Lyell's travels in North America ; I approve entirely of it, as far as it goes, but it is odd that he should have so entirely omitted all allusion to the University question, on which Lyell has bestowed so much attention. Read 7 scenes of " Die Deutsche Kleinstädter" with Fanny. Edward came in to see us in the evening, and gave us information about Peel's scheme of commercial reform, which has just been expounded to the House of Commons. The sliding scale of duties on foreign corn is to be maintained for a while, but at a much diminished rate, so that when wheat sells at 54s. a quarter in this country, the duty is to be only 4s., and after a few years (Edward was not certain whether 3 or 4) the duty is to cease entirely, and foreign corn to come in absolutely free. In the meantime, the duties on a great many other articles are either to be entirely repealed or much reduced, and in particular, *maize* is to be free from all duty, which is considered very important, as in years of scarcity it will be valuable as human food, and at other times for fattening cattle.

Edward stayed with us till one o'clock, talking very pleasantly on all sorts of subjects.

January 28th.

Peel's scheme of commercial reform expounded last night, is indeed of immense extent and

1846. importance. It sweeps away entirely the principle of *Protection*. The import duties on all kinds of food, animal and vegetable are to be entirely abolished, either immediately, or at most at the end of three years, and those on foreign manufactured articles are much reduced, so as no longer to serve for a protection to our own manufacturers, but only to contribute to the revenue. Besides all this, there are numerous other important changes, *ostensibly* designed as a kind of compensation to the landed interest : the most material are those which relate to parochial settlements, to the maintenance of prisoners in jails, to medical relief in the poor-law unions, to the education of children in the work-houses, and all these appear, as far as I can yet judge, to be very beneficial alterations. In short, I think Peel's scheme on the whole a grand and states-manlike one. But it is so vast and manifold, that Parliament may have plenty of work for all this session in debating the details of it. The resistance of the agricultural members appears likely to be vigorous, if we may judge from the vehemence with which several of them have already expressed themselves in the House. The serious debate on the corn question is however put off till next Monday week. It is very odd that not one member of the league has yet expressed any opinion on this proposition: but they will be idiots if they quarrel with it. Whether the continuance of the duty for three years longer will be any real advantage to the landed interest is, I think, much more doubtful.

I read about 40 pages of Blanco White to my wife

and finished Edward's essay on the Topography of 1846. Rome, which is very good.

———

This month has been a most singular one in point of weather, mild to an extraordinary degree, often even warm, but excessively damp, with a great prevalence of rain and fog. This day however has been very fine.

———

A very pleasant party at Bedford Place. Owen, Murchison, Dr. Fitton, Mr. Hallam, Sir William Symonds and his son Captain Symonds, at dinner, and a great number more in the evening. I had the advantage of much talk with Owen, who was as usual delightful. Having mentioned the Aurocks or Wild Bull of Lithuania, the great rarity with which the British Museum has been lately enriched by the Emperor of Russia, Owen told me that the bones of three distinct species of Ox are found in the post-pliocene deposits of this country, together with those of the Mammoth and Rhinoceros, and that it is probable that all the three continued to exist to within the historical period. One was the Aurocks, identical with that still existing in the forests of Lithuania, and apparently the *Bison* of the Latin classics. This is double the weight of our ordinary

1845. tame bulls.   The second equally or even more gigantic, was the Bos primi-genius, which Owen believes to be the *Urus* of Cæsar and other Roman writers.   It had much larger horns than the Aurocks, but no mane.   It appears that both these formidable animals existed in the forests of Germany, and probably also of Britain in the time of the Roman Empire, and were exhibited in the amphitheatre, that they were gradually exterminated by advancing civilization, but at what period is uncertain.   Walter Scott, in his fine poetical description of the Wild Bull of Scotland (in Cadyow Castle,) appears to have mixed up the characteristics of the wild white cattle of Chillingham (which Owen believes to be of the same species as our common domestic cattle), with traditions respecting the Urus or the Bison.   The third species was a much smaller one, the original stock in Owen's opinion, of the small Welsh and Highland cattle. Our larger breeds of tame cattle are derived, he thinks from the Italian breed, introduced by the Romans, and this perhaps ultimately from the Brahmin Bull and Cow of India.   Owen told me that the Antarctic voyage was productive of much fewer interesting novelties in the zoological than in the botanical department ; that not much was discovered which had not been found in Cook's voyage, and nothing of very marked interest or importance.

I had also much pleasant conversation with Dr. Falconer about the Punjaub and the Sikhs ; with Mr. Tremenheere upon politics ; and with

John Phillimore upon law abuses and law reforms. 1846. It seems to be expected that Peel will have a majority of upwards of 100 in the House of Commons on his Corn Bill.

February 3rd.

Finished the 2nd volume of the "Life of Blanco White."

February 4th.

I completed my 37th year. God grant that I may make a good use of the years that may remain to me.

Meeting of the Geological Society :—a long paper by Mr. Cumming on the tertiary strata of the Isle of Man. These deposits, which were described with great minuteness, appear to be chiefly Post-Pleiocene, or what used to be called diluvial, gravel, boulders, and drift of various kinds.

February 6th.

Finished reading with my wife, Kotzebue's very amusing little play of "Die Deutsche Kleinstadter."

Went to the Royal Institution, for which Mr. Horner had given me a ticket, and heard an interesting lecture frem Owen on the Geographical Distribution of Extinct Mammalia. He began by pointing out the leading facts relative to the geographical distribution of the recent forms of that great class of animals :—that the true Ant-eaters, covered with hair, were peculiar to South America, were represented in Southern Africa by a form

I

1846. essentially distinct, though calculated for living on
the same food, and in India and tropical Africa by a
still more different form, that of the Scaly Ant-eaters
or Pangolins ; that the Sloths and Armadillos, and
the prehensile-tailed Monkeys, were characteristic of
South America ;—that the marsupials predominated
almost exclusively in Australia ; that the largest car-
nivora and pachydermata belonged to Asia and Africa,
the American forms of those sub-classes being smaller
as well as less numerous ; and so forth.   He then
proceeded to the extinct forms :—observed, that the
oldest formation in which any traces of *mammiferous*
life appeared was the Stonesfield Slate, in which
have been found the bones of a small Marsupial
quadruped.   No remains of mammiferous animals in
the strata above this, till we come to the Eocene
tertiary beds, in which occur the remarkable
Pachyderms of the Paris basin, the Palaeotherium,
Anoplotherium, etc.   The bones of these in the
freshwater formation of the Isle of Wight as well as
in that of the Paris basin.   In the Meiocene series,
teeth of a Mastodon found in the Crag of Norwich,
the same species of Mastodon occurring more
abundantly in deposits of the same age on the
Continent.   Lastly, in the Post-Pleiocene or newest
tertiary deposits,—gravel, silt, shell-marl, etc,—are
the remains of numerous mammalia still existing, of
others which have become extinct in this island
within the historical period (such as the Beaver,
the Wolf, the Bear, three distinct species of wild
Ox), and of others again of which we have no
historical trace, (the Elephant, Rhinoceros, Hippo-

potamus, Hyæna, the great Irish Deer and another species of Deer equally gigantic, but with horns very differently formed, etc). Owen pointed out the proofs that all these animals had really lived in our islands,—that their bones had not been drifted hither,—and the geological evidence that our islands had not been severed from the Continent at the time when these races existed. He then pointed out the remarkable analogies between the geographical distribution of recent mammalia and of those of the post-Pleiocene age. For example, as the Sloths and Armadillos are at present confined to South America, so their huge extinct representatives, the Mylodon, Megatherium and Glyptodon have been discovered only in that country ;—as South America possesses the largest rodent animal at present existing (the Capybara), so the largest of extinct Rodents, the Toxodon, is South American. So likewise, the bones of many extinct species and genera of Marsupials, allied to the Kangaroos and the Wombats, are found in caves in Australia, which is the native country of their recent analogies. But there seem to be strong exceptions to this general rule, in particular, the abundance of elephantoid quadrupeds (Mastodons) in America, where nothing analogous exists at present. I hear that Murchison is to be knighted, and that this intention was communicated to him in a letter from Sir Robert Peel much more flattering than the offer itself. He has already obtained, what has long been his great object, permission to wear the Order given him by the Czar.

Went in the morning, by appointment, to Mr.
Pratt's chambers, to see his fossils and minerals
from the coal mines lately opened in Asturias. The
fossil plants that he has as yet procured are not
numerous in point of species, though they occur, he
says, in great abundance. I saw five different forms
of Lepidodendrons, two of them identical, I think,
with common British species; two or three
Sigillariae in such a state as to be undefinable; the
ordinary form of Stigmaria, a few detached leaflets
of a curious little Fern, like one obtained by Lyell
from Nova Scotia, a very small fragment of another
Fern; some Calamites and various indeterminable
fragments of stems.—In strata intercalated between
the beds of coal, are abundance of fossil shells and
corals, very similar in their aspect to those of the
Silurian system, but which Mr. Pratt tells me, are
Devonian. Red hæmatetic iron ore in enormous
quantity accompanies this coal formation. Mr.
Pratt told me of one bed 60 feet thick, and several
miles in extent, of pure unmixed hæmatite. Cinnabar
also occurs plentifully, both in the sandstones
accompanying the coal, and even in the coal itself,
not in regular veins of any considerable extent,
but rather irregularly disseminated, several thousand
pounds weight of it have already been procured from
these coal mines; it is sometimes accompanied by
small globules of native mercury, and very often by
large quantities of the beautiful red sulphuret of arsenic
(realgar). Native copper is also found, and an ore
of antimony containing silver.

Mr. Pratt tells me that considerable improvement 1846. has taken place of late in the social condition of Spain, and he thinks there is now a better prospect than there has yet been of the developement of her internal resources.

February 10th.

Read Bürgers' Leonora with my wife.

February 12th.

Went with Fanny to the British Institution, having been there by myself the day before. The only pictures that strike me as particularly good, are—The Breton Conscript leaving his home, by Goodall, and Danby's "Grave of the Excommunicated" (it should rather be of the Suicide).

The Shipwreck (from Shakespeare's Tempest) by Danby, is likewise rather striking. Three pictures of Etty's, two of which (a Woman carried off by a Pirate, and a Woman bathing), are good specimens of his style.

February 13th.

We bought Corda's "Beiträge zur Flora der Vorwelt," for £2 17. Began to read Undine.

Began to study Corda's Beitrage. I have made sufficient progress in the knowledge of German to find much less difficulty than at first in making out whatever I want for my scientific objects.

At Babbage's party this evening, the celebrated Charles Dickens was pointed out to me. His appearance, I must say, is not much in his favor, though he has certainly a clever countenance.

I am told that a decision on the Corn question is not expected before next Friday. The debate began on Monday last, the 9th. Most of the speaking hitherto has been on the side of the Protectionists, who seem inclined to make up in the number of their speeches, for what they want in eloquence and argument.

February 15th.

We went to see the curious old Church of St. Bartholomew the Great, in Smithfield. It is one of the most ancient in London, and contains a great deal of pure Norman architecture of the best style.

February 18th

General Evans was elected for Westminster, by a large majority in opposition to Captain Rous. I strolled down to the polling places at Covent Garden and Trafalgar Square, but all seemed to be going on

very quietly, even tamely, and there was no appear- 1846.
ance of excitement.

I heard at Sir Edward Codrington's, where we
dined, that there is some degree of anxiety touching
the result of the debate on Peel's commercial scheme
and that it is not thought likely that the majority
for Government will be nearly so great as was at
first expected.

Began to read the third volume of Thier's
"Histoire Consulat et de l'Empire," of which I read
the first two volumes in December. It is very
entertaining.

Anniversary meeting of the Geological Society,
Mr. Horner presiding. Wollaston medal and
proceeds of Wollaston Fund awarded to Mr.
Lonsdale. An admirable address from the President. Dinner at the "Crown and Anchor," and
much *toastifying* and *speechifying* as usual. I was
fortunate in sitting next to John Moore, whom I
like particularly.

Sir Roderick Murchison was there, adorned with
the star of his Russian Order.

Conybeare appeared for the first time as Dean of
Llandaff. The affair was fatiguing, like all public
dinners, but went off very well.

February 23rd.

Finished the introductory part of Corda's "Beitrage" which contains much curious information.

The newspapers contain many details of the dreadful battle, or series of battles which has been fought in India, between our army and the Sikhs, the list of killed and wounded is frightful. Between fifty and sixty officers killed, and among them the heroic Sir Robert Sale and another Major-General. The ultimate result however is a victory on our side. Dr. Falconer told me, some time ago, that he was sure the Sikhs would fight well, and that their artillery was particularly formidable, and so it has proved. He says that the Indian artillerymen, as well in our service as in that of the native powers, have quite a religious feeling towards the guns which they are used to work, that they kneel down to them and embrace them, and adorn them with garlands of flowers on particular days, and would rather die than desert them. The terrible loss sustained by our army on this occasion seems to have been mainly caused by the Sikh's artillery.

February 25th.

Meeting of the Geological Society. I attended at the Council for the first time, having been elected into that body at the Anniversary, and we had a long debate on the Library regulations. At the General Meeting was read a long paper by Mr. Prestwich, on the tertiary formations of the Isle of Wight, which was considered valuable by those most conversant with that particular branch of geology.

Went to the Royal Institution to hear Forbes'
lecture on the geological causes which may have
influenced the present distribution of plants and
animals in the British Islands. It was very
interesting. In the main it was a developement of
the same views which he stated to the British
Association at Cambridge, and which have been
mentioned with high approbation by Dr. Joseph
Hooker in his " Flora Antarctica."

Forbes considers the vegetation of our islands as
made up of five distinct Floras derived from
different regions, and at different geological epochs,
partly mingled together, yet still preserving on the
whole a distinct aspect.

The prevalent Flora of the plains of England,
especially of the Eastern and Midland Counties, is
what he calls the Germanic Flora, all the characteristic species being plants of the Middle and North
of Germany, Belgium and Holland. This is, in
the British Islands, the most widely spread of
all the five Floras, mingling more or less with the
other four, thinning out as we proceed westward,
and thus indicating its Easterly origin.

Its date is referred by Forbes to the post tertiary
period, or that immediately preceding the era of
man, when (as there are geological proofs it is said)
there existed lands connecting the east of England
with Belgium and Holland.

The second Flora, a very well marked one, com-
prises the Alpine plants of the mountains of Scotland,
the Cumbrian region and Wales, all of which are

1846. likewise Scandinavian species, and are supposed by
Forbes to have migrated from thence. He refers the
period of this migration with very great probability
to the glacial epoch, when the peaks only of our
mountains appeared above the icy sea, forming
rocky islands, and when the climate was so much
colder that these Scandinavian plants could flourish
at a low level in our latitudes, as they now do in
Iceland. The communication with Scandinavia
may have been formed by chains of islands since
submerged, or the plants may have been brought by
floating icebergs.

The third Flora, much less distinctly marked, is
that of the south-east of England, chiefly developed
in the chalk districts, and agreeing with that of the
north of France. There is no doubt that England
and France were formerly united in that part where
the Straits of Dover now exist ; but this Flora does
not seem very well distinguished from the Germanic,
nor do I understand why the Professor throws back
its introduction to an earlier date.

Fourth comes the Flora of the Channel Isles,
Devonshire and Cornwall, and the South Coast of
Ireland, connected very evidently with that of the
west coast of France, and pointing to a time when
Boulogne was joined to the south-west of England,
either by continuous land, or by a chain of islands.
This state of things is supposed by the lecturer to
have existed in the older Pliocene period, or the
close of the Miocene.

Lastly comes the peculiar group of plants inhabiting the mountains of the west of Ireland, which

clearly appear to be derived from the North of Spain.
Forbes supposes that this small and very local
Flora is a relic of so early a period as the Miocene,
and that there then existed a tract of land connecting
Spain with Ireland, and extending across the present
Bay of Biscay. He observed that the situation of
the Miocene strata in Asia Minor, where they had
been upheaved to the height of 6,000 feet above the
sea level, as well as various facts in other localities,
clearly proved the occurrence of disturbances fully
sufficient to account for the disappearance of the
supposed land to the westward of our now existing
Europe. He even conjectures that this Miocene
land may have extended as far west as the Azores,
and that its western boundary may be indicated by
the remarkable and well-known band of Gulf-weed
or Sargasso, which is almost stationary in a
particular part of the Atlantic.

Professor Forbes afterwards proceeded to touch
slightly on the distribution of animals, in accordance with these views and mentioned a very
interesting and striking instance of the verification
of theoretical conclusions.

He had conjectured that marine animals of more
northern or Arctic forms would be found in the
deepest parts of our seas, just as arctic plants are
found on the tops of our mountains. In his
expedition to the Shetland Islands, last summer, he
and his associates dredged some very deep parts of
the sea near those islands, where there are known to
be, as it were, deep pits or hollows in the sea-bottom,
and they found in fact, that these depths were

1846. inhabited by animals of truly Arctic species, such as are found in comparatively shallow water near the northern extremity of Norway.

February 28th.

The Ministerial plan of Free Trade has passed through the first step of its ordeal, the proposition for going into committee on the resolutions having been carried last night, by a majority of 97, after a twelve night's debate.

March 3rd.

We spent some time in the zoological gardens, very pleasantly. The weather beautiful and the season wonderfully forward ; the lilacs and weeping willows bursting into leaf, the almond trees in full blossom and everything looking thoroughly like spring.

March 11th.

Meeting of the Geological Society.—Sir Roderick Murchison presided, in the absence of Mr. Horner. Darwin and Forbes were at the council, and I had some talk with them. I asked Forbes why he considered the Flora of the south-eastern or cretaceous districts of England to be more ancient than the prevailing or Germanic Flora, as he had stated in his lecture ? He said that he was himself very doubtful on that point, and that it was the part of his theory on which he had least made up his mind. At the evening meeting was read a paper by Captain Vicary, on the geology of that part of Beloochistan which was traversed by Sir Charles Napier's army in the expedition against the robber tribes of the hills. The principal rock appears to be a limestone particularly characterized by the abundance of numulites ; this is overlaid in many places by a ferruginous gravel, in which are found great quantities of fossil wood and bones of large quadrupeds ; these bones being penetrated by oxide of iron like those found in the Sewalik Hills. —Murchison stated that from 15 to 20 of the fossil shells and echinida sent home by Captain Vicary had been ascertained to be identical with species formerly collected in Cutch, and described in the Society's transactions.

He read part of a very characteristic letter from Sir Charles Napier, giving a most spirited and striking account of the aspect of the country. The Dean of Westminster (Buckland) and Sir H. de la Beche, expressed their approbation of the paper. Mr. Hamilton and Mr. Forbes spoke of the similarity between the geology of this tract and that of Asia Minor ; and Forbes made some interesting observations on the vast formation of numulite limestone *(scaglia)*, which appears to range from the south of Spain through Italy, Greece, Asia Minor, Syria, and Persia, even to the Indus, and which he

1846. suspects to be equivalent in point of time, not merely to the cretaceous system, but to the *whole* of the secondary series of our countries.

The next paper read was a short notice, by Dr. Taggart, of some supposed casts of foot-prints of birds, lately found at Hastings.   This gave rise to a very entertaining discussion, Murchison and Buckland bandying jokes with each other, with great good humour.

Mantell gave his opinion that it was somewhat doubtful whether the objects in question were foot-marks at all, but that if they were so, they were more probably the tracks of reptiles than of birds; perhaps indeed they might be those of the Iguanodon.

March 12th.

Met Mr. Brown at Bedford Place.

March 17th.

I attended the first of Owen's course of Lectures on the Osteology of the Vertebrated Animals, at the College of Surgeons, Owen himself having sent me a ticket.   After some introductory observations on the value of the study of Anatomy, he gave us a general view of the characters common to the whole class of Vertebratæ, and then pointed out the 1846. principal modifications characteristic of each of its leading divisions — Fishes, Reptiles, Birds, and Mammals.

March 18.

Re-visited the Museum of Economic Geology, with my wife, and Katharine and Joanna. We accidentally fell in with Sir H. De la Beche, who volunteered to be our guide and showman, and took a great deal of pains to point out and explain to us the most interesting departments of the collection.

March 23rd.

Went with Fanny, Susan, and Katharine, to the British Museum, called on Mr. Brown, and saw some magnificent Vellozias—entire trunks, with their roots and leaves, which have just been sent from Brazil. In the old stems, the small proportion which the real stem bears to the mass of leaf-sheaths and air roots by which it is encased, and which form the apparent bark, is still more remarkable than in the small branch which I brought home with me. We spent some time in the print room, looking over beautiful engravings.

1846.   March 24th.

I re-visited the Museum of Economic Geology, by
myself, and spent some time in studying the
collection of metallic ores, which is very instructive.
Afterwards went to Owen's lecture, at the College of
Surgeons. This lecture related principally to the
different parts of which a vertebra may be considered
to be composed, and the modifications of the
vertebræ in the subkingdom of Fishes, beginning
with the Branchiostoma or Lancelet, the lowest or
most simply organized of Vertebrata,—then proceeding to the Lamprey, which is but a little higher
in the scale of organization,—then to the Sharks
and other cartilaginous Fishes, and the anomolous
Lepidosiren, which though a real Fish, has been
taken for a reptile.

March 25th.

In the evening was at the meeting of the
Geological Society, when a very good paper by
Charles Darwin, on the Geology of the Falkland
Islands was read, also Lyell's notice on the Coalfield
of Alabama ; and there was also a good discussion.

March ᴘ7th.

My wife and I went down to Sandgate on the
coast of Kent.

We spent there three days of great enjoyment and returned this day to town.

The accounts of the brilliant victories gained over the Sikhs by Sir Harry Smith at Aliwal, and by Sir Hugh Gough at Sobraon, coming in rapid succession engrossed for a time all one's attention, but in the account of this last and decisive victory, came the painful news that Charles Lyell's brother was severely wounded. The very next day, April 2nd, we learned that the short and splendid career in India had been terminated by a peace as honourable to the moderation of our Indian Government, as the battles were to the valour of our men.

Monckton Milnes, whom I met at a ball at Mrs. Nightingale's, said that he thought Cobden was really in earnest in his wish to dissolve the Anti-Corn-law League, though the underlings who owe all their importance to the agitation, will of course do all they can to keep it up.

A very pleasant small party at Bedford Place, where I had the great satisfaction of becoming acquainted with Dr. Joseph Hooker, whose "Flora Antarctica," has raised him to the very first place in the new generation of botanists. He is a very pleasing young man, of mild and gentle manners,

1846. unaffectedly modest, and at the same time communicative, and ready to talk on scientific subjects.
He has just commenced the study of fossil plants,
having been appointed botanist to the Ordnance
Survey. I had much pleasant talk with him. We
spoke of the rich and remarkable flora of Lord
Auckland's Islands, which he was the first to
examine : he said that, though in many respects
peculiar, the vegetation of those islands, has on the
whole a decided similarity to that of New Zealand,
but with the striking difference of the entire absence
of Coniferæ, which in New Zealand are particularly
numerous. Neither do the peculiar beeches of
Fuegia extend to Lord Auckland's Islands. He
confirmed Darwin's account of the excessive density
and gloom of the Fuegian woods composed principally of those beeches, and said that at the sea
level the beeches are large trees, 60 or 70 feet high,
but they diminish in height as you ascend the
mountains, till towards the upper limit of their
range, they become mere shrubs, with widely
extended branches, so matted together, and so
rigid, that it is easier to walk over their
tops than to make way between them. I
mentioned Lyell's account of the Fir trees on
Mount Washington, in New Hampshire, which
dwindle in the upper part of the ascent, to procum-
bent shrubs, matting the ground, and not rising
higher than the club-mosses with which they were
intermixed. Dr. Hooker said that there was a
remarkable difference in this respect between the
Pines of North America and of Europe, the

European kinds always retaining the character of 1846. trees wherever they can grow at all. He told me that a species of true Pinus has been discovered in Borneo, the first that has ever been found South of the equator. Speaking of Ferns, Dr. Hooker said he was satisfied that their geographical range is in very many instances very extensive, far more frequently than has been generally supposed, and that forms occurring in different countries have been described as distinct species by persons observing them only in this or that country, when a comparison of a large series of specimens from various regions proves beyond a doubt that they are the same. This is the case still more with the Lycopodia ; for instance, forms undistinguishable from Lycopodium clavatum are found in almost every part of the world. This wide diffusion of cryptogamous plants, as Dr. Hooker observed, is easily accounted for, as their seeds or spores are so very minute and light, that they may be carried to any distance by the winds, and quantities of them may very probably be constantly floating in the air, germinating only when they meet with a favourable soil and climate. Thus Lycopodium cernuum, which is a common plant throughout the tropical regions, grows near the hot springs in the Azores, and no where else beyond the northern tropics.

April 6th.

I heard to-day, at the Geological Society, that

K 2

1846. Colonel Sabine conjectures the extraordinary mild-
ness of this last winter to be occasioned by the Gulf
Stream extending farther to the north-east than
usual. He says that this was ascertained to be the
case in the winter of 1821-22, which was likewise
remarkably mild.

Sir William Symonds, speaking of the Martello
towers on the coast, told me that in the late war, at
San Fiorenzo in Corsica, a Martello tower with a
single gun, disabled two 74-gun ships, which
battered it for a considerable time without effect;
and they were so much damaged as to be obliged to
return to England to be repaired.

I have been engaged, at Mr. Horner's suggestion,
in writing an analysis of " Corda's Beitrage," for
the Journal of the Geological Society, and have
finished it all but the Ferns.

April 8th.

Meeting of the Geological Society. A long and
important paper by Sir Roderick Murchison, on the
distribution of drift and erratic blocks in Scandinavia. Sedgwick, De la Beche, and Buckland
took part in the ensuing discussion, which was
rather entertaining.

April 9th.

Fanny and I went to Graves's print shop, to see
a picture, said to be by Raphael, which has been
lately brought to this country and is on sale there.
It is a portrait, apparently of some ecclesiastic

of high rank, and is certainly a noble picture, 1846. whether really by Raphael or not ; it reminds me in some degree of the portrait of Cæsar Borgia, in the Borghese Palace at Rome.

The Lyells sent us from Kinnordy, a copy of Captain Henry Lyell's interesting letter written two days after the battle of Sobraon. Happily his wounds do not appear to be dangerous.

Went with Fanny to the British Museum and looked over a portfolio of Callot's wonderful etchings.

Meeting of the Geological Society, Mr. Morris read an interesting paper on the structure of Terebratulæ, pointing out the concurrence of certain structural characters, by which that extensive and important genus can be divided into natural groups. Edward Forbes spoke highly of the value of this paper. Then came two short papers, one by Mr. Binney, on a specimen discovered in the Duckinfield colliery, and now in the Manchester museum, which is supposed to prove that Stigmaria is really the root of a tree. The other by a Mr. Brown of Sydney, Cape Breton, on some upright fossil trees, in the carboniferous strata of that country, which

1846. appear to adopt the same vein of the nature of Stigmaria. A very good discussion ensued on this
question, in which the Dean of Westminster, myself,
Sir Henry de la Beche, and Dr. Mantell took the
chief part. Buckland declared himself satisfied
with the evidence of the *root* theory, and Mantell
was quite vehement on the same side, not allowing
that there was even the least room for doubt.

To De la Beche and myself, on the contrary, the
evidence did not appear conclusive or unquestion-
able.

De la Beche mentioned several instances observed
by himself in the Welsh coal fields, in which
deceptive juxta-positions occurred that might lead
to the belief of connexions between Stigmaria and
the erect stems in the strata above them. He also
suggested that essentially different things might be
confounded under the name of Stigmaria. The
Dean mentioned the rhizome of the Nymphæa, as
exhibiting an articulated insertion of the fibres
analogous to what occurs in Stigmaria, and seemed
to think that this latter was of the same nature;
if so, it would be a rhizoma, or subterranean stem,
not properly a root ; but in the recent vegetable
kingdom, as far as I am aware, such *rhizomata*
belong only to herbaceous plants.

April 24th.

Went with Fanny, Susan and Leonora, to the
Botanic Garden, at Kew, and spent two delightful
hours there. The collection appears to be in even
finer condition than when I visited it last year. It

is hardly possible to imagine plants more flourishing, 1846. in more luxuriant health and vigour than they are in all the houses here. I was especially delighted with the Ferns, which are very numerous and of exquisite beauty. Among them I noticed particularly a magnificent plant of Hemitelia horrida, with immense fronds in full fructification. Marattia elegans, from Norfolk island, almost equally grand ; Cyathea arborea, another splendid plant, but not in a fertile state; Neottopteris (the old Asplenium Nidus) with its grand circle of tall, stiff, shining swordlike fronds, covered with fructification; the Drynaria Quercifolia, from the Indian Islands, of which the primary fronds have a striking similarity both in form and venation to large Oak leaves (and by the way, I here observed that the two kinds of fronds were in several instances united in one, the base of the frond having the obtusely sinuated oak-leaf character, while the upper and larger part was more deeply cut into large acuminated lobes). The Platycerium aleicorne and grande, two very striking Ferns ; a Davallia (pyxidata ?) from Australia, with ascending or almost erect stems, exactly inter-mediate in character, between the creeping stem so common in tropical Ferns, and the more robust trunk of the arborescent kinds; a very beautiful little Gymnogramma, with the back of its fronds covered with dense powder, of a brilliant orpiment-yellow colour ; and fine flourishing plants of many of my old Brazilian friends, such as Didymoch-laena sinuosa, Gymnogramma tomentosa, Lygodium volubile, etc.

1846.   The temperature of this Fern-house was at present 77 degrees, and the air extremely moist. In company with the Ferns were a vast variety of Orchideous plants, in fine condition, some of them in blossom, particularly the rare Phalænopsis amabilis, with its flowers very much resembling large white butterflies; a very beautiful purple Cattleya, and the curious little Masdevallia, which I gathered formerly at Gonga Soco.

In the New Zealand house are fine young trees of the famous Kowdie pine, Dammara Australis, and of the Dacrydium cupressimum, the latter very graceful; several species of Podocarpus; the Phyllocladus very flourishing. The Banksias and Dryandrias in the Australian house are in great beauty, and there are young plants of several of the Cape Proteaceæ and of the rare Guevina Avellana from Chili.

Certainly these gardens appear to me more deficient in Cape plants than in those of any other considerable botanical region; and so indeed Dr. Hooker told me they were.

There is a grand new Palm house now building, intended to rival the famous one at Chatsworth.

We were unlucky in not meeting with Sir William Hooker on this occasion, but I was introduced by my wife to Mr. John Smith,* whose arrangement of ferns has given him a high place in the botanical world.

April 29th.

Went to both the water-colour exhibitions of the

---

* One of the gardeners.

Old and the New Society,—in each of which I saw 1846.
some pretty things. Then strolled down to the new
suspension bridge near Hungerford market, from
whence, the day being bright and clear, I enjoyed
very agreeable views both up and down the river.
Waterloo bridge, Somerset house, and St. Paul's on
the one hand,—Westminster bridge, the Abbey, and
the new Houses of Parliament on the other, were
seen to great advantage, — the number of steam
boats, continually passing up and down the river, is
really astonishing.

April 30th.

Dr. Falconer tells me of a very remarkable and
interesting discovery which has lately been made,
and which is not yet published. Major Rawlinson
and Mr. Norris of the Asiatic Society, have succeeded in deciphering the famous arrow-headed
inscriptions of Babylon and Persepolis, and find
them to contain historical records of great interest
and importance, tending altogether (as far as they
have yet been read), to confirm the statements
of Herodotus. One inscription, he says, is a record
in the name of Darius (Hystaspes) of the principal
events of that monarch's reign.

It is an interesting and pleasing fact, that at the
last general examination of the London University
three of the prizes were gained by Hindoos.

LETTERS.

16, Hart Street,
April 29th, 1846.

My Dear Mary,

1846. It was immediately after I had finished my last letter to your husband that we were shocked by the painful news of Captain Lyell's being severely wounded in the last of those terrible battles on the Sutlej.

Happily the last accounts of him are very satisfactory, and his wounds do not appear likely to produce any permanent injury ; but we felt very much for the pain which you must both have suffered, receiving only the statement taken from the Gazette, and having to wait for the next mail from England before your anxieties could be at all relieved. I am delighted that the war is over, and that so much moderation has been shown in dictating the terms of peace. Dr. Falconer indeed thinks that the Sikhs have been let off much too easy, but for my part I quite approve of moderation towards the vanquished, and though I dare say the Sikh troops will continue to bear no good will towards us, yet as they have been obliged to give up almost all their artillery, they are not likely again to be formidable enemies.

The time of our departure from London is now drawing very near (we have settled to go on the 9th of May), and I confess I look forward to it with

pleasure, for I find so long a stay in town disagrees 1846.
with me; my strength is failing very much, and
I long for pure air and exhilarating rambles. We
are going to my father's place in Wales, and do
not expect to return to Suffolk before the end of
June, by which time I heartily hope you will be
safely arrived in England. Do not suppose that
because I am growing tired of London, I am at
all the less grateful for your kindness in lending
us your house; this winter in town has been of great
advantage to us in many respects, and particularly
in improving my acquaintance with several persons
whom I highly value, but I feel that my health does
not bear it well. I am delighted to hear such satisfactory and interesting accounts of your travels,
though rather surprised at your being so well pleased
with New Orleans, for all former accounts had given
me the idea that it was a very disagreeable place;
however very much depends on the colouring which
one's own mind gives to objects, and I think you
differ from all previous travellers in America in the
rose-coloured medium through which you view everything.

Really I do not think there ever were two people
so perfectly formed for travellers as you and Charles
Lyell. How happy we shall be to see you again!
and how much we shall have to talk over. You
really must come down to Mildenhall. Since my
last letter I have had the great satisfaction of
becoming acquainted with Dr. Joseph Hooker, who
(as you perhaps know), has been appointed to the
botanical department of the Ordnance Survey, in

1846. connexion with the Museum of Economic Geology, and has just begun to turn his attention to the study of fossil botany. I like him extremely, he is so modest, and at the same time so communicative and full of knowledge and talent. His "Flora Antarctica" is an admirable book, and gives him I think the very first place among the botanists of the new generation.

We went down to Kew the other day, and spent two delightful hours in the gardens, though the weather was very unlike what it was at our first memorable visit there, in company with you. The various houses were in prodigious beauty, the plants looking in the highest possible state of health and perfectly bewildering from their profusion and variety; but above all I was delighted with the ferns. The collection is certainly poorer in Cape plants than in those of any other important botanical region. By the way I was much pleased to hear from Dr. Hooker that when he touched at the Cape, he derived great assistance in his botanical re-searches from my notes which were inserted in the "Journal of Botany."

Your father has lately made a most noble present to the Geological Society of his copy of Agassiz's Poissons Fossiles. There was indeed one copy of it already in the library, but as it is to be kept there constantly for reference, it is very desirable to have another for circulation. This example of liberality was very handsomely followed by several members of the council : John Moore, Murchison, Greenough, Mr. Daniel Sharpe and Mr. Bowerbank,—so that the library will be enriched with duplicate copies of 1846 several of the most valuable works.

We had a very good discussion last Wednesday at the Society, on Stigmaria, in which I took a part. Some more evidence as to its *radical* nature was brought before us, both from Manchester and Nova Scotia, but not quite satisfactory yet. Sir Henry de la Beche is still incredulous, and I am doubtful.

Of great news there is none, as far as I am aware, and small news I must leave to your sisters to communicate.

Fanny is well, and very busy and active. My father and his party are probably still at Rome, but on their way homewards.

Ever your very affectionate brother,
C. J. F. BUNBURY.

JOURNAL.

May 2nd.

A visit from Sir Alexander Johnstone, who was very pleasant, and told me many interesting things about India. In particular that in the south of India, no very great distance from Madras, and in the former dominions of the Nabob of Arcot, are the ruins of a Hindoo university, which many centuries before the Christian era enjoyed great influence and reputation ; and that among and near these ruins have been found many very curious inscriptions relating to mathematical and astronomical researches, and proving the mathematical sciences to

1846. have been cultivated with good success in India in those very early times.

Sir Alexander Johnstone's mother, when she resided in that part of India, spent much money and employed great numbers of people in making researches and excavations among those ruins, and found many curious things; in particular, an inscription recording that a solemn sacrifice of thanksgiving had been offered up to the deities on the discovery of some important mathematical or astronomical rule. Her researches were carried on upon so large a scale that they excited the suspicions of the Nabob of Arcot, who imagined that her object was to discover hidden treasure, and actually applied to the Governor of Madras, to prevent this English lady carrying away treasure which ought to be *his* property. By the Governor's advice, the lady sent a confidential person, well known to the Nabob, and well acquainted with the languages, habits and ideas of the natives, to explain to the Nabob the nature and object of her pursuits. The Nabob cured of his suspicions, was so much struck with her wisdom, that he bestowed on her the whole property in the ruins, and the tract of jungle immediately surrounding them, as a free gift to her and her descendants ; and Sir Alexander says that that tract still actually belongs to him. It is very remarkable, as a contrast to Hindoo manners at present, that it is recorded, Sir Alexander says, that men and women were educated together at this University, and without any distinction of castes, not even Pariahs being excluded.

Again at Kew, by myself, in the omnibus, and
spent two hours there, devoting most part of this
time to a more particular study of some tribes of
plants, as in the former visit my attention had been
almost bewildered by the multiplicity of objects.
The gardens are in great beauty; vegetation in the
open air has made much progress even since we
were there ten days ago; the azaleas in profuse
blossom as well as the lilacs. I devoted my attention
principally to the Ferns, Cycadeæ, and Palms.
The hothouse which contains the Ferns and
Orchideæ is a most beautiful sight. Among the
Ferns, besides those mentioned in my journal of the
24th, I especially noticed the beautiful Adiantum
trapeziforme, Blechnum Corcovadense, a very fine
and stately plant,—Darea rhizophylla, most delicate
and graceful,—Pteris (Cassebeeria) pedata,—
Doryopteris palmata, which is so like in form to the
last as to be scarcely distinguishable, but darker
in colour, and essentially distinguished by its
reticulated veins,—and several species of Cyrto-
phlebium, Goniophlebium, Goniopteris, Phlebodium,
and most of Mr. Smith's new genera.

I observed that many of these Ferns seed themselves spontaneously, and in great profusion, on the
damp soil, and among the Sphagnum, and on the
lumps of wood on which the Orchideous plants are
cultivated.—In the same house are three different
kinds of Nepenthes,—the distillatoria very flourishing.
They have made a new entrance to the gardens,
by a very handsome ornamented iron gate.

Went with Fanny to the Royal Academy
Exhibition, which opened on the 4th. The
exhibition this year is a very good one, better than
usual, I think, several of our best painters appearing
in great force. Landseer has some excellent
pictures : his " Stag at Bay " is a grand design, full
of spirit and character, but less agreeable in colouring
I think than is usual with him. " Refreshment " is
charmingly painted, it represents a grey pony,
tired and soiled with travel, feeding before a cottage
on the ascent of the Italian Alps, and a couple of
noble dogs of the St. Bernard breed lying near him ;
the landscape delightful. Then he has two companion or rather contrasted pictures called " Time
of War" and "Time of Peace." In the former we see
a magnificent black war horse, wounded and struggling
on the ground, amidst corpses of men and shattered
arms, and clouds of smoke, and the ruins of fallen
buildings ; in the latter, a tranquil sunny scene on
the coast near Folkestone, sheep and goats grazing,
and lambs playing about a rusty dismounted cannon.
The contrasted ideas of the two designs are very
well sustained and happily expressed.

Eastlake exhibits only one small picture of a nun
visited by her sister, very pretty and carefully
painted, but by no means striking ; however it
would probably gain upon one if one had time to
study it carefully, and especially if seen by itself.
There is something subdued and tranquil, if not
cold in this artist's works, which makes them suffer
particularly from the glare and glitter of surrounding

pictures.—Maclise's " Ordeal by Touch " (a suspected murderer compelled to touch the corpse of his
victim, from which the blood is bursting out), is
a very striking picture, and highly characteristic of
the artist's manner, with all his bold and luxuriant
imagination, his power of expression, his tendency
to theatrical exaggeration, his crowds of figures, his
hard outlines and cold metallic colouring. Of Eddis,
who exhibited such noble attempts in the grand style
of art at the two preceding years' exhibitions, there
is nothing here but a rather insignificant picture of a
Gipsy Fortune-teller. Etty has several things in
his usual style, of which his " Sea-bather" appears
to me the most successful in his peculiar way, the
flesh wonderfully natural and life-like. "Circe with
the Syrens and Naiades," is not so good, and in his
"Judgment of Paris," the heads of the three god-
desses are as common-place and ignoble as in
Rubens' picture, while their bodies have less appearance of real living flesh and blood than he
usually gives. There is a great sameness, both
in the forms and faces of his beauties,—it seems
as if he always painted from one model.

There is a cleverly painted little picture by Mulready, "The Choosing of the Wedding Gown (from
the Vicar of Wakefield), surprisingly rich in colour,
finished like a Dutch painting, and with a good deal
of character.

Of landscapes there are several worth mentioning,
especially a delightful sea coast view, seen by the
light of early morning, by Collins ; and Stanfield's
view of the Ponte Rotto at Rome. I say nothing

of Turner's dazzling and bewildering pictures (of which there are several here), because on a first view, at least, they are utterly unintelligible to me.

In the evening, the meeting of the Geological Society :—a paper by Mr. Prestwich, on a Section of the Wealden Strata, exposed by a railway cutting near Tunbridge Wells ; and a Paper by Lyell on " The Tertiary Deposits of Alabama."   I had some pleasant talk afterwards with John Moore and with Mr. Daniel Sharpe.

JOURNAL.

CHAPTER II.

Left London at half-past eight, went by railway to 1846 Wolverhampton, and thence in a post-chaise to Mr. Whitmore's, Dudmaston Hall, about four miles from Bridgenorth.

At Dudmeston. The house is good and the place a very pleasant one. From the library windows one looks down on a slope, adorned with prettily laid out beds of flowers, to a fine piece of water, beyond which rises a smooth green hill, variegated with trees, and over the ridge of this are seen the dark tops of the Clee hills. At a very short distance from the house, a path leads one through a very pretty wooded dell, called "The Dingle," and thence through a pleasant wood covering the side of a steep hill, immediately above the Severn, the winding course of which, glittering in long bright reaches through the rich, green meadows, are seen to great advantage from several points in the walk.

1846.   These woods are full of beautiful spring flowers ;
the lovely Wood Forget-me-not, the Blue-bell, Red
Campion, Wood Anemone, Wild Strawberry, Woodsorrel, and yellow Dead-nettle, carpet the ground ;
the Foxglove is beginning to throw up its stately
spires of blossom, and abundance of Ferns are just
now unrolling their delicate curled-up leaves.

The country hereabouts is very pretty and pleasant,
—rich, verdant, and luxuriant, well-wooded, finely
varied with hill and dale, and not without occasional
bolder features.   At this season too it appears to the
highest advantage from the exquisitely rich and
brilliant green of the grass fields, and the variety
of soft and delicate tints in the young foliage of the
woods.

Dudmaston Hall is in the parish of Quatt, about
four miles south of Bridgenorth.   Quatford, which is
on the road between these two places, is a pretty
village with a handsome church placed in a fine
commanding situation.   The left bank of the river
here is steep and high, particularly at one point,
close to Quatford, where it rises into an abrupt
knoll, presenting to the river a bold picturesque
escarpment of red sandstone, with shrubs and
vigorous young trees springing from the crevices
of the rock.   This knoll is enclosed on the land-side
by a deep and broad trench, extending round
all that part of it which is not precipitous.   It is
called a Danish camp, and tradition says it was the
last stronghold of that people in this part of the
country, when they were retreating before Alfred.

The great building material hereabouts is the

deep red sandstone of the country, which when 1846. not quite *raw* and fresh, has a good effect, and is said to be very tolerably durable.

I have nowhere seen so great a number of fine old yew trees as in this neighbourhood. They are not however usually in situations where one can easily suppose them to be indigenous. There are two of vast size in the dingle at Dudmaston.

The Severn here is navigable, yet not so deep that it may be forded in many places, as I am told. Its bottom is pebbly, its water clear, and the current very strong, so that it must be hard work for boats to ascend it. Very great quantities of charcoal are made in this neighbourhood, and there are many forges on the banks of the river, which frequently have a picturesque effect.

The rock of this district is the Lower New Red Sandstone of most English geologists, answering to the Rothe Todte-Liegende of the Germans, and belonging to Murchison's Permian system.

Mr. Whitmore is a great farmer, and does much good in the country, by his example, and by his exertions in the cause of education, and his attention to the condition of the people. He has not literary or scientific tastes, but is a specimen of the best sort of *practical* men—a rational steady liberal in politics, and most meritorious for the efforts he made in behalf of free trade, long before it became fashionable, or was taken up by any Ministry.

———

May 14th.

From Dudmaston by Bridgenorth, Much Wenlock

1846. and Buildwas Abbey, to Shrewsbury, 27 miles. Stopped at Wenlock to see the ruins of the Abbey church, which are fine, though not in a striking situation. A considerable part of the transept remains standing, of very graceful and beautiful architecture,—Buildwas Abbey, which stands amidst low rich meadows, on the banks of the Severn. Its remains are considerable, but less beautiful than those of Wenlock Abbey. The nave remains nearly entire, and appears to be of the Norman time, having simple and very massive round pillars (not clustered) and arches very slightly pointed.

Cross the Severn by an iron bridge, very near to these ruins, and look up the valley to Coalbrook Dale. The valley of the Severn from hence to near Shrewesbury very beautiful, the river winds most gracefully, and the country of it is very rich, finely wooded, and exquisitely green. Passed nearly under the Wreken ; it is a curious isolated ridge, its direction from south-west to north-east, or thereabouts, and from most points of view, its appearance is somewhat like a camel's hump, but seen endwise, it appears quite a bold peak.

A few miles before we come to Shrewsbury, a piece of Roman wall is seen in a field immediately on the left of the road. It is constructed in the style usual in works of a somewhat late period of the empire, of thin courses of the flat Roman bricks, alternating with much thicker courses of squared pieces of sandstone. The masonry appears to be good, and and its character is truly Roman.

Although the Town (Uriconium) of which this is

a remnant, is not celebrated in history, yet it is 1846.
always interesting to meet with the visible traces of
that wonderful nation.

Shrewsbury is a fine and picturesque old town,
apparently almost surrounded by the Severn. It
contains a remarkable number of quaint looking old
houses of timber, black and white, with high-
pointed gables towards the street, and has altogether
a much more antique air, than most English towns,
we visited the Abbey Church, remarkable for a very
large and rich window, in the perpendicular style,
at its west end; and for several old monuments;—
the fine old Market Hall,—St. Mary's, a beautiful
Church,—and the Infirmary, from the terrace of
which there is an extensive and very agreeable
view.

May 13th.

We left Shrewsbury very early in the morning,
by the Aberystwith mail, and went on by Welshpool
and Llanfair to Mallwyd, from whence we took a
chaise to Abergwynant. The weather was lovely,
and the country delightful. We reached Abergwynant at about half-past four,—the distance from
Shrewsbury said to be about 56 miles.

May 19th.

At Abergwynant.

This house is situated in a small valley or hollow
of a somewhat elliptical form, bounded on the south

1846. by two very steep and stony mountains of considerable height, which are offsets or spurs of Cader Idris,—and on the North by lower, but very bold and rough hills, shaggy with wood, separating the valley from the estuary of the Mawddach. The mixture of grey rock, heath and young wood, gives a singular beauty to these knolls, which much resemble those about the Trosachs Pass near Loch Katrine.—The direction of the little valley is nearly east and west ; the greatest part of it is occupied by rich green meadows, with swampy spots here and there. A swift and exquisitely clear stream, coming down from Cader Idris, and passing between the two mountains first mentioned, through a beautifully wooded ravine, flows obliquely across the valley, and escaping between two of the shaggy knolls, falls into the estuary. Looking up the ravine through which this brook comes down, between the two steep mountains, we see the glorious cliffs of Cader Idris forming a grand termination to the view, on that side. From the hills we catch fine views of the estuary, which from some points, when the tide is in, has quite the appearance of a lake, so closely do the mountains seem to hem it in on all sides, and occasionally bright glimpses are seen of the open sea beyond. The scenery is truly delightful.

May 24th.

Vegetation in the moist and rocky woods here is extremely luxuriant. The heath grows taller and stronger than I have ever seen it elsewhere. Ferns 1846. flourish in exceeding profusion and beauty, with a luxuriance which I have never seen surpassed except in tropical regions. Mosses also are in exuberant abundance, and of the finest growth, covering the rocks, the damp ground, and the roots of trees, with thick soft cushions of the richest and most varied tints of green, yellow, brown, and even purple, and adding greatly to the beauty of the woods. The delicate modest blossoms of the Woodsorrel, and the bright little golden stars of the wood Moneywort (Lysimachia nemorum) peep out everywhere among the moss; the shaded margins of the rivulet are decorated with the brilliantly white flowers of the Wild Garlic, and the Bilberry, with its bright green leaves and pretty waxy flowers, grows luxuriantly among the Fern and the rocks. The Foxglove and the Heath, which at a later period will be the great ornaments of the place, are not yet in blossom; and I see no trace of the wood Forget-me-not, so abundant and so beautiful at Dudmaston.

The woods continue in their full richness of growth to the very edge of the estuary, and it is rather a striking sight for a botanist to see the woodland Ferns and Mosses, and Heaths and Bilberry bushes, growing within a few feet of Fucus vesiculosus and other sea weeds.

May 26th.

We dined at Mr. Reveley's, about a mile on this side of Dolgelly.

1846.   Among the distant mountains seen from hence,
Mr. Reveley pointed out to us Arran Mowddy, which
is said to be higher than Cader Idris, but is not of
so fine a form.

The Dolgelly river (the Ynnion) is in this part a
bright, clear tranquil stream, of moderate size,
winding through bright green meadows, along a
narrow valley; the tide seems not to affect it here.
A little lower down, it is joined by another river
coming in from the north, and then the valley
suddenly widens into a broad and singularly level
expanse, very much resembling the bed of a lake.
This flat surface is marshy, covered with vegetation,
and the river winds through it in a strangely tortuous
and irregular course.   It terminates about 2½ miles
below Dolgelly, where the river becomes a true
estuary.

May 27th.

We went to visit Mr. and Mrs. Edwards at
Dolcerrau, two miles or more on the other side of
Dolgelly.   They took us up the beautiful glen of
the Clwydog, a tributary of the Ynnion.   It is a deep
and wild ravine, richly wooded, at the bottom of
which the impetuous stream forces its way through
a succession of magnificent rocky chasms, leaping
and foaming from rock to rock in a long series of
beautiful little falls; the grand and picturesquely broken masses of rock, covered with moss, fern and wild
shrubs, in some places quite overhang, and conceal
the torrent; and old trees extend their feathering

branches almost from bank to bank.—It is a 1846
beautiful and romantic scene.

In this dell grows abundance of the globe flower,
which I have not seen at Abergwynant.

May 28th.

A delicious day. We went up our mountain
Fridd Defad, and enjoyed most charming views of
our peaceful little green valley, the tufted and
shaggy knolls which bound it, the mountains on the
other side stretching away to the roots of the
gigantic Cader Idris, the windings of the estuary,
the mountain barrier beyond it, and the blue glittering sea. We sat down on the fresh elastic mossy
turf of the mountain, beneath the fantastic crags of
its ridge, and enjoyed to the full the scenery, the
sunshine, and the exhilarating air.

At sunset, the whole crest of Cader Idris was
coloured with an indescribably rich glow of rosy
purple, equal to anything I ever saw in Italy, or the
Alps. It was the more striking by contrast with the
deep green of the woods, between which we looked
up to it, and which were already in shade.

May 29th.

Walked to Arthog, which is a little more than six
miles from Dolgelly, near the mouth of the Mawddach, and full in view of Barmouth. Here the
mountains bend away to the south, and leave, on
this side of the estuary, a low marshy plain of some

1846. miles extent, perfectly level, except that some detached rocky hummocks or knolls, rise out of it like islands.

Doubtless they were islands at no very distant period when the sea covered these marshy flats.

This was a perfectly Italian day, not a cloud in the sky.

May 30th.

Another lovely day, and the second anniversary of my happy wedding day, an occasion which I shall bless as long as I live.

We went in a car to Barmouth, returning in the evening. The whole drive from Dolgelly to Barmouth is a constant succession of beauties.

The valley for two or three miles below Dolgelly is delightfully green and smiling, and the low hills which immediately bound it, richly wooded; farther down, where the river widens into an estuary, and the valley dilates in proportion, the scenery is of a grander character; whichever way we turn, we see mountains most beautiful in their forms, colouring and grouping; some almost impending over the road, others faintly seen in the glowing haze of the distance. The variety in the appearance of the hills and mountains is indeed very striking; some are covered with smooth green turf, some brown with heath, others craggy and bare, or overspread with loose, grey stones; nor are their forms less various.

LETTERS.

My Dear Leonora,

I thank you very much for your letter, 1846. which I received on our arrival here, and for the very entertaining and satisfactory account contained in it of the Horticultural fête. The beautiful Wisteria seems, I fear, to be feeling the approaches of old age, which is a great pity. With respect to Ferns, I do not know of anything like a good *iutroduction* to the study of them ; you will nowhere find more information about them than in Hooker's three Works, the "Icones," "Genera," and "Species Filicum."

We are enjoying this most lovely place, in spite of the wet and stormy weather which has come on, and we have been rambling and scrambling famously. The fresh green of the young foliage makes it particularly charming at this time of year. The Ferns and Mosses are most beautiful, and in extraordinary profusion and luxuriance, and I am busy collecting for you as well as for ourselves. I have already made out a list of 75 species growing here, but only 25 of them are flowering plants, and among these I have not yet found any thing that can be called rare. How delightful it is to

1846. know that the Lyells are safe again in the civilized
parts of America, and that all the perils of their
western journey are well over.   I think Mary writes
as if she had had quite enough this time, of such
rough travelling.   The last news from America
seems pacific, and at any rate it is quite clear that
if war does ensue (which I do not believe), it will not
be for some time to come, and the Lyells will have
plenty of time to get home.

We are both quite well ; we enjoyed very much
our stay at Mr. Whitmore's and our subsequent
journey, and were much pleased with Shropshire,
which for an inland county, and one without grand
monntain scenery, is as pretty a country as any I
have seen in England.

We were fortunate in having delicious weather
for our journey Dudmaston hither, and for our first
day here, so that Fanny's first impressions were as
favourable as could be, and she is as much charmed
with the place as I could wish.

Best love to mamma and our sisters,

Ever your very affectionate brother,

C. J. F. BUNBURY.

JOURNAL.

June 5th.

There are some interesting birds here.   Herons
are numerous on the shores of the estuary.   I have
seldom gone down thither without seeing some of

these noble birds stalking along the shore, or passing 1846. over it with their heavy flapping flight. I have also repeatedly seen cormorants on the shore, and sheldrakes once or twice. The kite, which has become a rare bird in most parts of England, is still found here; I have more than once seen it sailing above our hills, with that peculiarly smooth graceful, gliding motion which is characteristic of it. Mr. Hugh Reveley tells me that the peregrine falcon is to be found on Cader Idris, and that the bittern still exists in the low marshy flats along the Mawddach, between this place and Dolgelly. Owls seem to be numerous in our rocky woods; we hear them hooting every night.

The other evening, as my wife and I were walking along the top of the hill between our house and the estuary, a goatsucker got up from among the heath, and flew for some way along the path before us, with a peculiarly low and undulating flight.

We are much struck with the prodigious number of the large dark-brown wood ant. Every path in the woods, and even in the garden, swarms with them, and it is most curious to observe their indefatigable activity, industry, and perseverance, and the strength they exert in carrying bodies much larger than themselves.

During some days, there were great numbers of a small kind of cockchafer about our garden, and whenever any of these fell to the ground, as they very often did, a swarm of ants would fasten upon them, overpower them, and drag them off in spite of their struggles. The nests of these ants

1846. are very large, loose heaps of bits of dead wood, dry leaves, grass-stalks, and other such materials.

Weather for the last week brilliantly clear and very hot. Vegetation accordingly has made rapid progress, and the woods, which at the time of our arrival had still a spring-like appearance, are now clothed in the deep green of full summer. The Foxglove is everywhere displaying its beautiful spires of blossom among the woods and rocks.

LETTERS.

Abergwynant, near Dolgelly,
June 8th, 1846.

My Dear Lyell,

I write this note to congratulate you on your safe arrival (as I hope and trust) in England, for before another week is past I trust you will be at home. I really cannot express to you how happy it will make me to see you and Mary again. How much we shall have to talk over!—Fanny and I must be in London on Saturday or Sunday next, and *I* must go down to Bury on Tuesday, about a plaguy piece of magisterial business, but I trust to have, at any rate a glimpse of you in the day or two that I shall be in town, and I hope to return from Bury on Wednesday evening to find you still in Hart Street. But I look forward with especial hope to seeing you both at Mildenhall in the course of the Summer or Autumn, and having a comfortable

long visit from you, there are no two people in the 1846.
world whose society I value more than yours.

Fanny and I have been above three weeks at this
most delightful place, and have enjoyed it most
thoroughly, being favoured nearly all the time with the
finest weather possible. I do not know a more lovely
or more enjoyable spot. We are, both of us, quite in
love with it, and shall tear ourselves from it with great
unwillingness. I do hope some time we shall meet
you and Mary here ; I am sure you would be pleased
with the place, and it would be such a pleasure to us
to ramble over it with you. The geology is interesting, the slates being intersected and intermixed in a
complicated manner with trap rocks, which assume
a great variety of appearances.

I shall be very curious to see your Alabama coal
plants. I have nothing more to say but to repeat
the assurances of my love to both of you.

I am ever

Your very affectionate brother-in-law,

C. J. F. BUNBURY.

JOURNAL.

June 28th.

I wish to begin this new volume of my Journal with
an expression of my humble and hearty thanks to
Almighty God, for the many and great blessings He
has bestowed on me. Most gratefully do I acknowledge the inestimable advantages I enjoy in the
counsels and examples of my excellent father, in the

M

1846. good education I have received ; in the affection of
my admirable wife, whose fine moral sense and
unfailing love of truth and virtue, are so well calculated to correct the faults of my character and to
guide and support me in temptation ; and in the
attachment of several valuable friends, above all of
such a friend (Charles Lyell) as few men have ever
been blessed with.

I thank God also for the enjoyment of a healthy
constitution, and that I am placed in such a station
of life as to have ample leisure and opportunity for
intellectual pursuits, and to be free from the
harrassing care for daily subsistence.   I pray that I
may never in future misuse these advantages and
opportunities, that I may grow wiser and better as I
grow older, that I may struggle successfully against
the faults of an indolent temperament, that I may
contribute to the increase of human knowledge and
happiness, and that when my life is drawing near to its
end, I may not have to look back on the whole of it
with repentance and shame.

We left Abergwynant very reluctantly on the 12th
(June), had a pleasant journey on the top of the coach
to Chester, in company with Hugh Reveley ; slept at
Chester, and arrived in London the next day.   On
the 15th, in the evening, we had the great happiness
of seeing the Lyells arrive safe and well, from their
fatiguing, and in part dangerous travels in America.

I was obliged to go down to Bury, the following
day, as the election of a new Chief of the rural
police was coming on ; but I returned to London on
the 18th, and we did not leave it till the 25th.

During the 20th and 22nd (Saturday and Monday) 1846 I worked hard at examining and describing for Charles Lyell, the vegetable remains which he had collected in the Alabama coal field.

We returned to our home (Mildenhall) after a seven months' absence, on the 25th, and found it very pleasant and home-like, the garden very gay and pretty, and our favourite Skye most boisterously glad to see us.

All Friday and Saturday I was chiefly occupied in unpacking and arranging, but I had time to execute a little drawing for Lyell, and have sent it off to-day. It was very pleasant to find myself settled and at work again in my own old favourite museum.

My plans for the next two months are as follows; to prepare a memoir on Lyell's fossil plants from Richmond in Virginia; and to work up carefully any other materials that may be within my reach, in the way of fossil botany.

To finish " Corda's Beitrage," of which I read about half when I was in London, and to prepare an abstract of it for the Geological Journal; also to read Göppert's work on the fossil remains in Amber.

To examine and arrange my collections of Mosses and Lichens, of which a considerable part still remains in confusion.

To go on with the study of German.

To read Xenophon's Anabasis, his Hellenica and his Socrates, in Greek, or as much of them as I have time for.

To read Hallam's literature. I doubt whether I

1846. shall manage all these, as I must reckon upon many
interruptions, but I will try to work hard.

June 29th.

Rose at 8 ; read Corda from 10 to 11 ; then began
to examine and describe the fossil plants from
Virginia, which, together with drawing, occupied me
till 2.   After luncheon strolled about the garden
with my dear wife for some time, and afterwards
took a walk to the College plantation and " Bombay"
where I found the bogs almost dried up by the
heat of this extraordinary season.   Saw Ranunculus
Lingua in great abundance and beauty, and gathered
uncommonly large specimens of Veronica scuttelata.
Sky-larks and Wheat-ears very numerous.—In the
evening began to read Hallam's literature.

June 30th.

Rose at 8 ; read the newspaper which announces
that Peel's important bills for the repeal of the
Corn-laws, and for the alteration of the customs
have received the Royal assent,—and at the same
time that Peel has offered his resignation, though
his successor has not yet been sent for.   It is a
curious state of politics.   It is said also that a treaty
settling the Oregon question, has been agreed upon,
between the American Government, and our Minister
at Washington, but no particulars appear to
be known.

Read Corda from 10 to 11; he is a very tedious

and bad writer, though seemingly a very diligent 1846. and accurate observer. Examined and drew fossils from 11 till 2. Read the first chapter of Xenophon's Anabasis, and about half the second. Lounged about the garden and took no walk. Read Hallam in the evening.

July 1st.

Rose at 8. Read in the newspaper Peel's speech of Monday evening, announcing his resignation. It is in a fine, generous and manly spirit, and even touching; he declares that even if he had been successful in carrying the Irish Coercion Bill, he would have retired from the government, as he had irrevocably lost the confidence and support of his former party, and he could not bear to be Minister by sufferance of the Opposition. He very fairly and handsomely acknowledges that the honour of abolishing the Corn-laws is really due neither to him nor to Lord John, but to Cobden.—So the Whigs are again to attempt the government, but they are a weak party, and if they try to form a *merely* Whig ministry of the old materials, I think they will hardly last over one session of parliament. And then one does not clearly see what they are to stand upon, or what great measure remains to give them a chance of popularity. The question of Free Trade is settled; they must dissolve parliament, and upon what question are they to appeal to the people? If, indeed, they had courage to attempt a grand and comprehensive scheme of National Education, in

1846. defiance of the exclusive claims of the Church or of any sect, it would be a noble attempt, but would probably be fatal to them; they would however fall with more honour than on the former occasion.

It seems to be certain that the Americans have accepted the compromise proposed by our Government in the matter of the Oregon territory.

This is a very happy event.

I read Corda from 10 till 11; then went on with my drawing of the beautiful Neuropteris from Richmond in Virginia, and began to settle my unarranged Lichens.

July 2nd.

The Evening Mail of course gives Sir Robert Peel's speech much more fully than it was in the Bury paper in which I read it yesterday. I am very much pleased with it. In the course of it he stated the terms proposed by his Ministry to the American Government in relation to Oregon, and read an official letter, received by the last mail from Mr. Pakenham, announcing the acceptance of these terms by the Americans. So we may consider that threatening question, as happily settled. Peel, I think has adopted a far wiser and more dignified course in thus resigning at once, than if he had clung to office, as the Whigs did in 1840 and 1841.

The newspaper contains many details of the war which has broken out between our Government of the Cape, and the frontier Caffirs,—interesting to me,

because I know all the localities, and have a 1846.
lively recollection of the aspect of the country. The
Caffirs appear to fight better than they did in the last
war, and our troops who seem to be too few for
their work, and have to act in a most intricate
and difficult country, have met with some checks.—
Graham's Town itself is said to be in some danger.

I went on with my drawing, and with my arrangement of Lichens; read the third chapter of the
Anabasis; in the evening, finished the first chapter
of Hallam's literature.

July 3rd.

A very fine and hot day. Up a little before 8;
read a good spell of Corda; finished my drawing and
description of Neuropteris Linneæ Æfolia, and
made some progress in arranging my Lichens.
Read no Greek. Took a walk towards Eriswell in
company with Skye; gathered Hydrocharis Morsus
Ranæ in flower, and Phleum Boehmeri, but the
excessive drought of the season is very bad for
botany, especially in this flat and open country. A
heron flew over my head, when I was about half-a-
mile beyond the turnpike on the Eriswell road: these
fine birds are become rather rare in this part of
the country.

July 4th.

Very hot. I got up at quarter-past 8; went to
the Board of Guardians; there were but few cases,

1846. and none of any interest or importance. There are at present in the workhouse 7 men (none of them able-bodied), 7 women, and 24 children.—I examined and described some more of the fossil plants from Virginia, and arranged some of the Lichens collected long ago in Brazil and at the Cape. Read the fourth chapter of the Anabasis; and in the evening read part of the thirty-third chapter of Thirlwall's Greece, which gives the account of that expedition.

The evening mail says that Lord John Russell has already formed his Ministry, and gives a list which it says may be depended on. It is just the old Whig set, with no new men, even Charles Buller is omitted, which I think a great error. The Duke of Wellington however remains Commander-in-Chief. It is said that Lord John Russell made overtures on the one hand to Cobden, on the other to Lord Dalhousie, Lord Lincoln, and Mr. Sidney Herbert, but they all declined to join him,—Cobden merely on private grounds.

It is also said that Peel has promised not to engage in any *systematic* opposition to the new Ministry; so that, as there is no great question pressing for immediate settlement, the Whigs, though possessed of little real strength, may go on quietly enough for some time.

There are many more details of the Caffir War. It is evident that the object of the Caffirs, as on former occasions, is merely plunder, and especially to carry off cattle; and that in this object they have been to a great degree successful. They show no eagerness to fight, and whereever a spirited resistance

has been made to them, they have been repulsed;
but their great superiority in numbers, the extent
and defenceless nature of the frontier, their perfect
knowledge of the country, and their active and
stealthy habits, have enabled them to penetrate into
the Colony at many unguarded points, and to commit great ravages. Nearly all the Chiefs seem to
have joined in this invasion, even those of the Congo
tribe, who were before considered the most civilized
and friendly. It is evident that our regular troops
in that quarter are much too few for the extent of
frontier they have to guard, and the " burgher force "
from the other districts of the Colony had not come
up at the date of these accounts.

Thus ends a satisfactory, and I hope a well-spent
week.

July 11th.

During this week, I have not made much progress.
The greater part of it (from Monday to Thursday)
having been spent at Barton, with my father and
Lady Bunbury and Cecilia, who are just returned
from the Continent. Part of Friday was taken up
by the Petty Sessions, and part by the job of unpacking a large box of fossils, which is newly arrived
from Charles Lyell ; and in having new shelves put
up in my museum ; so that this day is the first I
have had clear for study, This morning, too, I had
an hour at the Board of Guardians ; but the greatest
part of the day I have been working hard at fossils.

1846. I have read no Greek or German this week, and at Barton, I read nothing of any importance.

My father tells me that the present King of Sardinia is doing great things in his Genoese and Piedmontese dominions, in the way of improvements and public works, and what is much more, has avowedly adopted a liberal course of policy, and allows himself to be proclaimed as the head of the *Italian* party. The position he has assumed in this respect is said to give great alarm to the Austrians; and now that a *Liberal* Pope has been elected (for such the new Pope is said to be), a great and happy change in the condition of Italy may be expected. The late Pope (Gregory XVI.) in his horror of all innovations, had especially set himsel against railways, so that a man who knew the Italians very well said to my father "The election of the next Pope will be a mere railroad question." His Holiness's subjects were most ardently desirous of these. innovations. Gregory XVI. was generally admitted to be a very good-natured, and even a benevolent man, but his extreme jealousy and dread of political movements rendered him a tyrant, and the Castle of St. Angelo and the other state prisons in his dominions were filled with respectable men, arrested on the merest suspicion.

The new coast road from Nice to Genoa, my father says is magnificent, rivalling the grandest of Napoleon's roads. The great tunnel through the mountain at the head of the Polcevera valley, for the railroad from Genoa to Turin, is well advanced; but

a far more gigantic work is said to be in contempla- 1846
tion—a tunnel through the Mont Cenis.

Lord John Russell's Ministry seems to be completed, and on the whole promises very fairly, though
there are some decidedly bad appointments. The
worst are. Sir John Hobhouse to the Board of
Control, and Mr. Fox Maule, Secretary at War.

Milner Gibson is Vice-president of the Board of
Trade, a compliment to the League. All is going on
wonderfully smoothly and quietly, scarcely any
symptom of opposition, anywhere, to the re-election
of the new Ministers. Party seems to be dead, or
rather asleep for awhile, but it will revive no
doubt.

July 19th.

In the past week, I have finished reading the first
book of Xenophon's Anabasis, and the 4th Chapter
of Hallam's Literature.

I have nearly completed my descriptions of the
fossil plants from Virginia, and have made some
progress in studying and arranging the Mosses and
Lichens which I collected long ago.

I have also read some more of Corda.

Yesterday was my dear wife's birthday. Mr. and
Mrs. Horner and Katharine arrived the day before.
Mr. Horner is not at all well. Lyell has sent me a
piece of a stem of Xanthorhœa, of which the
structure is very curious.

LETTERS.

My Dear Lyell,

1846.
    I am very much obliged to you for the stem
of Xanthorrhœa, which you sent me by Katharine ;
its structure is very curious and interesting, and
although it has already been described (in De
Candolle's Organographie), I am glad to have the
opportunity of examining it. I was very glad to
hear of your safe arrival at Kinnordy.

I have been busy unpacking and examining the
fossil plants from Cape Breton, sent me by Mr.
Brown, which arrived the day before yesterday, *per
varios casus, per tot discrimina verum.* It is a most
beautiful collection ; many of the Ferns in particular
are far finer than anything I had before ; several
new species, I think, and some specimens that throw
additional light on the structure of old species.
I hope to be able to draw up a paper on them,
which will also give me an opportunity of launching
into the subject of the geographical distribution of
plants.

There is a *Bechera,* precisely the same, I think, as
the one you collected in Alabama. I do possess the
sixth edition of your Principles, and I will proceed

to comment on the passage to which you have called
my attention. In the first place, it cannot be said
that the Flora of the carboniferous period appears
to have " consisted almost exclusively of large
vascular cryptogamic plants."

It is now admitted on all hands, that the Sigillarias
had no relation to Ferns. Brongniart is of opinion
that they belonged to the same class as the Cycadeæ
and Coniferæ (gymnogens of Lindley) while Göppert
maintains with perhaps equal plausibility, that they
were allied to the succulent Euphorbias. The
Calamites also are now referred by Brongniart to
the same class of gymnogens. At any rate the real
affinities of Sigillaria and Calamites are too uncertain to allow us to draw from them any safe
conclusions as to climate. Still the great numbers
of Ferns in the carboniferous Flora is very striking,
and I think it implies a mild, equable, and very
moist climate, but not necessarily an extremely hot
one; for New Zealand, which is by no means a hot
country, vies with any of the tropical islands in the
number of its Ferns (140 species of New Zealand
Ferns are now known). Another point of similarity
between the vegetation of New Zealand and that of
the carboniferous era, consists in the prevalence of
Coniferous trees; for the wood that is found with
its internal structure preserved, in the coal formation
is most commonly of the Coniferous structure, but
this does not bear directly on the question of
temperature.

I have alluded, in my paper, in the Journal of the
Geological Society, to Chiloe, which, as Darwin

1846. shows, enjoys very little summer heat, and yet has a
luxuriance of vegetation rivalling that of the torrid
zone.   I am not aware that Joseph Hooker has anywhere expressed an opinion as to the probable
climate of the carboniferous era, but in the London
Journal of Botany (v. 2. 328) he says that the more
he saw of the Ferns, the more he was convinced that
their geographical distribution chiefly depends on
a uniform and moist temperature, such as is generally
found in islands.   The Appallachian coal-fields
must have been a *thumping* big island by the
way!—I hope in spite of your new edition of the
Principles, you will have time to draw up a geological
paper on the Richmond coal-field, so that it may be
ready with my botanical one against November.
It is unlucky that Agassiz is so dilatory and uncertain,
and that Sir Philip Egerton has not been able to
examine your fishes.   I am very glad to hear from
Mr. Horner, that you have determined on publishing
an account of your second travels in America.   Pray
do not shrink from giving us *plenty* of geology and
natural history, I have no news to tell, and must
leave it to Fanny and Katharine to send bulletins
of health.   Pray give my kindest love to Mary,
and remember me to Mr. Lyell and your sisters.

Yours, very affectionately,

C. F. J. BUNBURY.

JOURNAL.

Horrible account in the newspapers of the fall of 1846. a barrack (blown down) at Loodiana, by which hundreds of men, women, and children, of the 50th regiment have been crushed or mutilated. There seems to have been criminal neglect on the part of the authorities, in allowing the troops to remain in so insecure a building.

Examined the anatomical structure of the creeping stem of the Buck-bean, and made a note of it.

Examined the anatomy of the stem of a Brazilian Aristolochia, and made a note of it.

Wrote another long letter to Charles Lyell, on botanical and geological subjects.

In regard to public affairs, what has most pleased me lately has been the news that the present Pope has released all the political prisoners, with whom the Castle of St. Angelo, and the other state prisons, had been filled by his predecessor.

Returned from a visit to Barton, where we had found

my father and Lady Bunbury in very good health and spirits. The last accounts of Cecilia are quite cheering.

My father showed us a curious thing, a large old elm tree in the park, which the other day burst asunder, and fell down spontaneously, without any shock of wind, or apparent external cause. It was perfectly rotten inside. We saw it in its fallen state exhibiting an extraordinary mass of ruin, as if it had been shattered to pieces by lightning.

In the Arboretum, Koelreuteria paniculata is in abundant blossom, (the first time it has flowered with us), and makes a very handsome appearance, it has large loose spreading pannicles of small yellow flowers, spotted with red at the base of the petals, and of a singular form; these are succeeded by large, triangular, inflated, reddish capsules. The Bignonia radicans is in fine bloom, and the Wisteria is producing a second crop of flowers, but the Tulip trees do not seem to have flowered this year. In the garden are some thriving young plants of the Tussack grass from the Falkland Islands, which Dr. Lindley sent to my father.

The newspapers are full of details of the extraordinary thunder storm which raged in London and its neighbourhood, on the first of the month, and did immense mischief. Here we had only an ordinary thunderstorm.

August 8th.

Since our return from Barton I have read the first and second chapter of the second book of the

Anabasis, and the first chapter of the second volume 1846
of Hallam's Literature.

I have been busy re-arranging my South American
herbarium, according to the system of Lindley's
" Vegetable Kingdom," which, though defective in
many points, seems to me, on the whole, the best
that has yet been proposed. This (I confess my
weakness) has somewhat drawn me away for the
present from the study of fossil plants. I have
however nearly completed the examination of the
collection sent me by Mr. Brown, from Cape
Breton.

A great deal of attention has lately been excited,
and much public feeling roused, on the subject of
military flogging, in consequence, more particularly
of the fate of a soldier in the 7th hussars, who died
at Hounslow barracks, in consequence, as it appears
of the effects of flagellation on an unsound constitution. The total abolition of this punishment
has been loudly demanded, but is resisted by the
Government, who, however, promise that no
punishment exceeding 50 lashes, shall in future be
inflicted, that the frequency of flogging shall be
restricted, and that the utmost attention shall
always be paid to the recommendations of the
medical officers.

It does not appear to me that the mere diminution
of the number of lashes is the reform most
needed. Fifty lashes may be more to one man than
200 to another, and at any rate, the degradation is
as great in the one case as in the other. My own
impression is, that flogging ought to be altogether

N

1846. abolished as a punishment for purely *military*
offences (except perhaps when troops are engaged
on active service in an enemy's country), and should
be reserved for real *crimes*, such as imply *moral
depravity*, not mere breaches of discipline. I am not
an advocate for the total disuse of this punishment;
on the contrary, I think that it might be employed
more frequently than it now is in ordinary criminal
justice, with very good effect. It is certain that the
fear of it has effectually put a stop to the mania for
shooting at our Queen.

The Committee of the House of Commons on
the Game Laws have made their report; I think it
will give satisfaction to nobody, except perhaps the
Squires.

_____

August 30th.

Yesterday I finished my abstract of Corda's
" Beiträge zur Flora der Vorwelt," to my great joy,
for it has been a very tedious job. I have now only
to write it out fair.

While we were at Barton, a week ago, the report
of the House of Common's Committee on the case
of the Andover Union, made its appearance. The
investigation had been a very long and laborious one.
The report not only exposes the monstrous abuses
and mis-management which had pervaded the whole
administration of the Poor Law in the Andover
Union, but pronounces a strong censure, though in
temperate language, on the conduct of the Poor Law
Commissioners—and with great reason.

It appears they had fallen into a most irregular

and lax habit of transacting business, that their 1846. conduct towards some of their assistant Commissioners has been harsh, arbitrary and unjust; that they have not encouraged those officers in detecting and bringing forward cases of abuse in the administration of the law; and that they most culpably endeavoured to stifle and hush up this Andover affair. I am amazed that the Commissioners have not resigned ; how any men with the feelings of gentlemen, and with a sense of shame, can persevere in holding office after incurring such a censure, and when such a storm of public indignation is directed against them, is incomprehensible. It is clear that a thorough revision of the whole system must take place early in the next session of Parliament, and the whole Commission must be re-organized, on a new footing ; with the exclusion, I think, of all who have hitherto been concerned in it. If the Whig Ministers make any attempt to screen the Commissioners, they will draw on themselves such a storm as they will not be able to stand against; they will fall, and deservedly. I have some fear that even the principles of the law itself may be swept away by the feeling excited against the Commissioners.

LETTER.

___

To Leonard Horner, Esq.

Mildenhall,
September 2nd, 1846.

My Dear Mr. Horner,

1846.

I am rejoiced to say that I am at length quit of that most tedious gentleman, Mr. Corda, and I send you my abstract of his book, which I hope will not be too bulky for insertion in the Journal. I hope it will be useful—at least, I know that I should have been very glad of such an abstract when I was beginning the study.

I am now setting to work at the Cape Breton fossils, which Mr. Brown sent me. My father has made me a generous present of a large cabinet, which will enable me to arrange my collection much more comfortably.

We have been quietly enjoying our garden and lawn, and Fanny's attention has been much absorbed by her poultry yard, which is a great source of interest and amusement to her; she has now a considerable stock of fowls, many of which are tame, even to impudence; and she has two little chickens which are especial pets, which follow her wherever she goes, and take food not only from her hand, but from her lips. Skye is jealous, and not without reason.

I am astonished that the Poor Law Commissioners 1846. do not resign, as I think the present Poor Law a good one in its principle, and useful if properly administered. I should be sorry if it were to be swept away by the public feeling against its mal-administrators ; but I think there is some risk of it.

I hope you will much enjoy your time at Southampton, and am sorry that I shall not " be there to " see."

Pray give my love to Mrs. Horner, and all my sisters, and to the Lyells, of whose safe arrival I was very glad to hear. Believe me,

Your affectionate son-in-law,

C. J. F. BUNBURY.

JOURNAL.

September 8th.

We returned from Barton, where we had spent a day and a half very pleasantly, Sir George and Lady Napier being there, and Fred Freeman. We had the comfort of seeing dear Cecilia quite restored to health ; Sir George Napier told several interesting anecdotes of what had occurred at the Cape in his time, but I cannot repeat them with anything like his spirit. This is one of them :—

Some English soldiers were tried at Cape Town, for a mutiny on board a convict ship ; one of them was condemned to be shot, and two to be transported

1846. for life. The governor, who suspected that the
man sentenced to death was not the worst of them,
delayed his execution, and caused the chaplain to
attend to him particularly. The man was silent,
sullen, and appeared to be thoroughly stubborn and
immovable; but after some time, the chaplain
reported that he thought he saw some slight
symptoms of softening in him. Then Sir George
himself visited the criminal repeatedly in prison,
talked kindly to him, and on one occasion, stayed
full two hours with him endeavouring to soften him ;
and thinking he perceived that the man was not
thoroughly depraved, and might be brought to
repentance, and being confirmed in this impression
by the chaplain, he determined to commute the
punishment, and himself told the prisoner of his
decision.   The latter looked astonished, but said
nothing, but the next time he was visited by the
chaplain, he told him that he could not resist the
governor's kindness,— that he was guilty, though
less so than the two who had been tried with him,
that he was heartily sorry for what he had done, and
would try to amend.   His dogged sullenness quite
vanished, and he appeared an altered character.
He was sent to Australia, and became quite conspicuous for his good conduct there.   If the sentence
of death had been executed, this man would doubtless have died hardened and impenitent.

Sir George also told us a pretty anecdote of a dog.
When the British Army was in the South of France,
after the battle of Toulouse, he and several other
officers visited the house of a gentleman who had a

very fine dog (a poodle, I think); this dog had been 1846
trained to receive food only when offered by the
right hand, and they all amused themselves with
testing his steadiness in this respect, and found that
he constantly refused to take bread from the left
hand. But when he came to Sir George, who
(having lost his right arm), of course offered the
bread with his left hand; the dog looked earnestly
at him, and accepted the bread. Then the other
officers tried to deceive him by disguising themselves
so as to appear to have lost the right arm, but
the dog's sagacity was not to be baffled, and he
steadily refused to take bread from the left hand,
except from the one who was really one-handed.

<hr>

September 11th.

Mr. and Mrs. Augustus Tharp called on us,
bringing with them my old friend Mary Anne
Houlton, who happened to be staying with them.
I was very glad to see her again, not having met for
many years,— indeed I think not since 1837,—
and she seemed glad to meet me. She is a very
pleasing person.

<hr>

September 13th.

I have been very much shocked and grieved by
hearing of the death of my dear friend John Napier,
who fell a victim to the cholera at Kurrachee, on the
7th of July. He was ill only six hours. The news
arrived yesterday. I was very intimate with him

1846. during the time we were at the Cape, and I loved
and valued him very highly, indeed, more than
almost any other man of my juniors that I have ever
known. He was a fine, noble, generous hearted
creature, frank, fiery, ardent and impetuous, full
of warm affections and generous feelings, with a refined and lofty sense of honour; ambitious, but with
no taint of unworthy sentiments in his ambition;
with talents that would have enabled him, I think,
to achieve distinction, if opportunities had occurred,
and with a strong inclination to improve his mind by
study, and to struggle against the prevailing idleness
of a military life. All this day his face has been
before me with singular vividness; and his voice has
seemed to be sounding in my ears, as in those
pleasant hours at the Cape, when we conversed
together on all subjects, grave and gay, in the un-
restrained confidence of intimate friendship.

I grieve for his father and sisters, and above
all for his poor young wife, who was on the eve
of her confinement when thus suddenly bereaved.
May God support her under this dreadful blow.

October 11th.

We returned yesterday from Hardwick, where
we spent two days and-a-half very pleasantly.
Sir Thomas and Lady Cullum were most kind and
hospitable. We met there Professor Henslow and
his daughters, and Mr. and Miss Mackinnon*.

* Afterwards Duchess of Grammont.

The house at Hardwick is decorated in too gaudy 1846. a style to my taste, and curiously crowded with trinkets and nicknacks, and what are called *objects of vertu*; so that Mr. Dawson Turner who visited there three years ago, compared it to Strawberry Hill. The library however, is a very pretty and comfortable room, and contains a good collection of books. The gardens and pleasure grounds are beautifully laid out, and kept in exquisite order; the evergreens remarkably fine, and in particular the hedges of holly, box and yew, which are of surprising extent and afford a delightful shelter from the fiercest winds. Many of the chesnut trees, planted by the present Sir Thomas, have grown into considerable trees, and have this year produced eatable fruit. The Quercus Ægilops, raised from an acorn, brought from the plain of Troy, is worth notice, it grows slowly, however, and has not yet borne any fruit; it is a scrubby-looking tree, with stiff, greyish, downy leaves, more deeply cut than they are represented in the Penny Cyclopædia, but less so than those of Quercus Cerris.

In the hot-house I saw Allamanda cathartica with its large bright yellow flowers, Passiflora racemosa, and another beautiful Passion Flower, with crimson blossoms, and the Stephanotis, all in high perfection, also a delicate little Passiflora with curious crescent shaped leaves, a Martynia, and a variety of Gesnerias.

The extent of glass is very great.

The beautiful Atalanta butterfly is remarkably abundant at present, and I have seen several of the

1846. Painted Lady, Cynthia Cardui. Mr. Henslow caught at Hitcham a Camberwell Beauty, Vanessa Antiopa, a great rarity, and he says that several of the Death's Head Sphinx have been taken in England this year.

October 12th.

Busy in arranging my fossils in the large new cabinet which has been put up in my museum.

October 13th.

Began to make a list of the Cape Breton fossils for Mr. Richard Brown, and resumed at the same time the writing of my paper on them, which had been for some time at a stand-still.

October 15th.

Finished the list to be sent to Mr. Richard Brown.

October 16th.

My servant John Norman left us, to my great regret ; he had been in my service since April 1837, and was a most excellent servant. The temptation of a railway clerk's place induced him to quit me.

October 20th.

Our dear friends, the Lyells, came to stay with us.

Edward arrived from London, and we had
also the Eagles and Mr. Hasted to dine and
sleep here.

October 28th.

The Lyells left us for London, to my very great
regret. Edward accompanied them. The week
they spent here, was a very happy one for me, as
indeed I never am in their company without deriving
both pleasure and instruction from it. The more I
know of them, the more I admire and love them
both, and as I have said before, I reckon their
friendship among the greatest blessings I enjoy.

I learned much from them in this visit respecting America. Their last journey in that country,
being chiefly through the southern and western
States, was much more adventurous, and exposed
them to much more of hardship and difficulty, than
the former one. The most laborious and unpleasant
part of the whole, Lyell said, was his expedition from
Tuscaloosa, in company with Professor Brumby, to
examine the coal mines, of which he was the first to
publish an account. The extension of railroads,
however, in every part of the Union, is very rapid, so
that every year regions are becoming easily accessible
which before could only be reached by a long,
painful and fatiguing journey. With respect to
slavery, the Lyells are strongly of opinion that its
evils (as regards the slaves themselves) are much
mitigated in the states they visited, though they
were struck with its injurious effects on the masters,

1846. and on the state of society generally. Happily, such
an improved state of public opinion has grown up even
in the Slave States, that any man who is notoriously
guilty of cruelty to his slaves is generally shunned and
scouted, and as there is no country in which public
opinion is so powerful, as in America, this may be a
better check than any positive law. In particular,
the Lyells affirm, there is now a very strong popular
feeling against the separation of members of the
same family, so that persons employed to sell slaves
by auction, have often been compelled, by the
expression of this popular feeling, to throw all the
members of a family into one lot, contrary to
the interest of their employer. I was also very glad
to hear that in some of the Slave States, the laws
prohibiting the education of slaves have become a
dead letter, so that many of the proprietors have
established schools for their slaves. Lyell however
admits that, while the state of things has thus
improved in the more settled countries, the old evils
have only moved on with the progress of colonization,
so that the abominations which formerly occurred in
Georgia and Carolina may now be taking place
in Texas.

One thing Lyell told me was quite new to me, and
struck me as very remarkable :—that the negroes
born in America are generally much superior in
intelligence to those brought from Africa, and that
the improvement appears to be progressive in the
succeeding generations ; he even thinks that the
physical characteristics of the negro race become
modified, without any intermixture of blood.

Lyell thinks as ill as any one can of the democratic 1846
party in America, but he believes there is so much
vitality, and so much of good in the nation, that the
constitution will ultimately, and perhaps soon, right
itself, without disruption or dangerous convulsions.
He is of opinion that the present war with Mexico,
if it runs on to a considerable length, and involves a
heavy expenditure must produce a real and important revolution in the nature of the government.
The customs duties, which at present form the
principal part of the national revenue (as distinct
from that of the several states), will very soon be
found insufficient for a war expenditure, and an
increase of the duties (which will probably be the
first plan the democrats will try) will rather injure
than assist their finances. Their credit is deservedly
so low, that they will be unable to obtain any aid
from loans; and thus they will necessarily be driven
to the resource of direct taxation. Now, the dislike
of the people to pay taxes for any but local purposes
is so great, that in order to enforce payment, it will
be necessary to strengthen the central government, and
arm it with unusual powers, and the war, originally
the work of the democratic party, may ultimately
lead to results very unfavourable to that party. A
war with England, as it would have caused a much
more serious derangement of the finances, would
more speedily have led to such a result.

Lyell thinks that some of the American Whigs
were secretly rather desirous of an English war,
looking upon it as the most likely remedy, though a
desperate one for the internal evils of their own

1846. Government. He is of opinion that if we were
forced into a war with America, our best policy would
be to avoid any invasion of their territory, to cripple
as much as possible their navy and their commerce,
to act merely on the defensive in Canada, and to
hold out every possible encouragement to any
State that might be inclined to conclude a separate
peace with us.

The number of German immigrants into the
United States, Lyell tells me, is immense, and they
seem in general to leave their own country from
feelings of political discontent rather than from
distress. They almost always fall into the hands of
the demagogues, and add to the strength of the
democratic and anti-English party. One of the great
evils in the present state of affairs in America seems
to be the haste with which these swarms of new
immigrants are admitted to all the privileges of
citizenship, before they can have acquired any sort
of knowledge of the real condition or wants or
interests of their adopted country.

Here is a saying of an American stage-coachman
almost worthy of Old Weller. Our friends were
travelling to Milledgeville, in Georgia, and the
railway to that place not being completed, they were
transferred to a stage coach, which carried its
passengers at the rate of about 4 miles an hour,
along an excessively bad road. The change was
naturally felt as disagreeable, and one of the passengers said to the driver—

"You cannot go quite so fast as the railroad
people?"

The coachman, nettled at this, replied—
" Yes, I can.   And faster too !"
" How do you make that out ?"
" Why, just you put one of those heavy engines on *this* road, and see whether it would go as fast as I do !   They go faster on their road, and I go faster on mine."

We accompanied the Lyells and Edward to Ely, and took a rapid view of the inside of that glorious cathedral, but could see scarcely anything of its exterior, owing to the fog.

The Dean* received us very kindly and cordially. Much progress has been made since I was last there, in the repair and renovation of the Cathedral, under his direction.

An extraordinarily thick and very cold fog prevailed without the least intermission through the whole of the 26th, 27th and 28th.   Winter seems to have come with one stride, after the longest and finest summer that I can distinctly remember to have known in England.

November 4th.

I heard of the death of my uncle, Henry Stephen Fox, at Washington.   It was very unexpected.   I am sorry for him, for he showed me great kindness when I was at Rio de Janeiro ; but his cold character, and eccentric and unsocial mode of life for many years past, and his entire disuse of correspondence, have so much estranged him from all his natural and

* Dr. Peacock.

1846. early friends, that I think there will be scarcely any one who will deeply feel his loss.

He was a very singular man, indolent to a degree that could hardly be exceeded, of great natural talents, remarkable wit, and great acquirements, though averse to severe studies; most agreeable and fascinating in conversation when he was in company with persons who suited him; very fond of the best society, but so intolerant of any but the very best that he preferred absolute solitude. The society of which Horace Walpole was the representative, seemed to be his beau-ideal, and in such a society he would undoubtedly have made a most brilliant figure. I do not think I ever knew a man, scarcely even Mr. Rogers, of a more exquisite taste in literature. His political opinions were such as might be expected from his family connexions and his character; in theory he was a strong liberal, almost a radical, but as all his tastes and habits were those of a refined and fastidious aristocrat, he would often express the strongest disgust at the practical effects of any approximation towards democracy.

November 6th.

Read the second and part of the third canto of the Orlando Furioso, having read the first just before the Lyells came to us.

I read in the evening, 82 stanzas of the " Orlando Furioso." Have been much engaged this week in examining and arranging an interesting set of North American plants, given me by Mary Lyell.

November 9th

We were obliged to go to Bury to inspect the hospital, in consequence of which the whole day was lost.

November 10th.

Began the 5th book of " Xenophon's Anabasis." Read the conclusion of the 4th canto of " Ariosto," and the whole of the 5th :—" The Story of Ginevra." The treacherous device of " Don John" in " Much Ado about Nothing " is borrowed directly or at second hand, from this.

November 11th.

Finished the rough sketch of my paper on the Coal-plants of Cape Breton.

Walked with my dear wife to the top of Barton Hill ; the day clear and bright.

Read the 6th canto of the " Orlando Furioso : "— Ruggiero's arrival in the Island of Alcina. The description of the monsters by which he is encountered is full of spirit and lively imagination.

November 12th.

Read the 7th canto of the "Orlando :"—"Alcina."

1846.

November 13th.

Read part of the 15th and 16th cantos of "Gierusalemme Liberata," to compare Tasso's "Armide " with Ariosto's " Alcina."

November 14th.

Read the 8th canto of the " Orlando Furioso."

November 21st.

A very fine day. Occupied all the morning, as I have been for some mornings past, in correcting and writing out my papers on the Cape Breton fossils.

We walked to Beck Row, and saw Mr. Manning* and several cottagers.

November 22nd.

A cold, stormy, wintry day. We did not go to Church. Hitherto we have been very regular in our attendance at Church during this summer and autumn. Read prayers at home with my wife, and in the evening a short sermon to the servants.

November 23rd.

Weather very wet and gloomy. I finished writing out my paper for the Geological Society, which has cost me a good deal of time and pains, but whether it is good for anything, I cannot tell.

November 27th.

We dined and slept at Elvedon, at Mr. Newton's.

* One of the Tenants.

We went up to London by the half-past twelve o'clock train from Thetford, and were most kindly welcomed by our friends in Bedford Place, and spent a very pleasant evening.

November 29th.

Went to see the Lyells in their new house in Harley Street, and had a very pleasant chat with Mary. Their house is a very good one, much superior to that in which they lived in Hart Street. Edward and Henry met us at dinner.

November 30th.

Dined with the Lyells, and spent a most agreeable evening. The party at dinner, besides ourselves consisted of Mr. Rogers, Mr. and Mrs. Milman, and Mrs. Marcet. Rogers appears a good deal broken, not in looks, for he has looked immensely old for many years past, but he has grown very deaf, shows less animation, and seems altogether to be failing. Milman is a very agreeable man ; his wife handsome and lady-like. After dinner came the Richard Napiers, Augusta and Miss Sherriff, Mr. Babbage, Sir Henry de la Beche, Sir Charles and Lady Fellowes, Mr. Bancroft (the Minister from the United States), and many others.

Sir H. de la Beche talked to me a long time : he is very desirous that I should undertake the office of Foreign Secretary of the Geological Society. I had

1846 much pleasant talk with the Richard Napiers, and
with several other persons.

December 1st.

The hearing of my father's tithe-suit commenced
in the Court of Queen's Bench, before the four judges.

December 2nd.

Meeting of the Geological Society.   My paper on
the Coal-plants of Cape Breton, was read, and was
well received, and what particularly pleased me was
the favourable opinion of it that Dr. Hooker expressed.   But the discussion on it was cut very
short, in order to make room for Daniel Sharpe's
paper on " Slaty Cleavage," which after all had to be
very much clipped, itself, that it might be finished
this evening.   Sharpe's paper, however, seemed, as
far as I could understand it, to be a curious and
important one.

December 3rd.

I went with Joanna to the British Museum, and
we took a rapid view of nearly the whole collection,
a pretty considerable miscellany !   Among the new
additions is the head of a fossil elephant from the
north of India, of astonishing size, with tusks *fourteen*
feet long ! This is one of Dr. Falconer's discoveries,
and he has very happily named it Elephas *Ganesa*,
a name peculiarly appropriate, as Ganesa is a
Hindoo deity with an elephant's head.   There is a
fine collection of fossil fish, which is in great part
arranged.   The shells appear to be a splendid
collection.   The department of antiquities, with the

exception of the Egyptian room, and that containing 1846. the Elgin marbles, is just now in great confusion, as the removal of the several collections of marbles to their places in the new rooms is in progress. We had a glimpse of the sculptures newly brought from Boodroom, and supposed to have belonged to the tomb of Mausolus. They represent combats between Greeks and Amazons, and strike me as being very spirited, but rude. They are much defaced.

In the afternoon I walked down to Cadogan Place, and saw Henry Napier. Two volumes of his "History of Florence," are now published, but he is very prematurely alarmed and annoyed at the slowness of its sale. It is not the sort of book that could reasonably be expected to "go off" very rapidly.

In my way, I looked at the much criticised statue of the Duke of Wellington on the arch at Hyde Park Corner, and I quite agree with those who think unfavorably of it. It appears to me an ugly statue in itself, independently of its position, and decidedly disproportioned to the arch ; too large for an accessory, and not large enough if the arch is to be considered merely as a pedestal. It had a strange and striking appearance as I looked up to it from Grosvenor Place in the dusk of the evening; the fog and the dim uncertain light making it seem much more huge than it did by daylight; it looked most grim and lowering and phantom-like.

December 4th.

Walked with Joanna to Harley Street, and had a most agreeable chat with the Lyells. Mary read

1846. me a letter from her friend Mr. Ticknor, of Boston, giving an account of a most remarkable and important discovery lately made by Dr. Jackson, a scientific chemist of that country :—the discovery, of a gas prepared from ether, which being inhaled, produces an absolute and total insensibility of some minutes' duration. The insensibility is so perfect, that the patient undergoes the most severe surgical operations, not only without pain, but without even the slightest muscular contraction ; in fact, he is for the time, in the condition of a corpse. Seventeen minutes is the longest time for which this insensibility has yet been made to last. The experiment has been tried in more than 200 cases in the hospitals of Boston under the eyes of the most scientific surgeons and physicians, without a single failure, and without any injurious consequences in any case. Mr. Ticknor's own daughter had two double teeth extracted while under the operation of this gas, without feeling the slightest pain. The discovery seems to be one of the most beneficial that can well be imagined.

I was very glad to hear also from Mr. Ticknor's letter, that the Whig party in America is gaining much ground, and has every prospect of a complete victory over the Democrats in the next elections.

Mary Lyell showed me a small but very choice collection of New England plants, given her by Mr. Oakes, the specimens beautifully preserved.* Dr. Hooker came to dinner with us at Bedford

---

* These she left to my husband. I now possess them. 1888, 12th October.

F. J. BUNBURY.

Place. I was exceedingly glad to meet him, and 1846 the impression he made on me was quite as agreeable as the first time. His manners are modest and gentle and very pleasing, and he is full of knowledge, which he is very ready to communicate. I was particularly gratified by the manner in which he spoke of my Cape Breton paper. *Laudari a laudato,* —to be appreciated by one whose own writings have raised him to so high a rank in the scientific world, is as good an encouragement to continue my efforts in that branch of study as I could desire. We talked of Forbes's paper (published that summer), on the origin of the different Floras which now compose that of Britain ; Dr. Hooker thinks it hasty and not in all points accurate. He does not admit Forbes's bold hypothesis respecting the Sargassa weed (that it marks the boundary of a submerged continent, a " Miocene land)," nor does he admit the necessity of such a hypothesis to explain the facts.

He believes that the floating Sargassa (sargassa bacciferum, etc.) are only altered and abnormal forms of species which grow on the coasts of South America and the West Indies, and that their remarkable accumulation in that particular part of the ocean is owing to the meeting of currents. He found similarly abnormal oceanic forms of some northern Fuci in the Atlantic, north of the zone of Sargassum.

Dr. Hooker mentioned a very remarkable instance of the effect of vegetation on climate. When the Island of St. Helena was first colonized, it was covered with wood and the climate was very moist ;

1846. the trees having all been destroyed, partly by the colonists and more by the goats which they introduced, the quantity of rain was diminished to a degree that proved very injurious, so that the inhabitants took to planting again as an antidote to the drought; an influence of the plantations on the climate is said to be already perceptible. Sir William Hooker's collection of plants occupies (his son told me), eleven rooms of his house. It is the finest private collection in the world, and scarcely surpassed by any public one, except that of the Jardin des Plantes at Paris. As for Baron De Lessert's, Dr. Hooker thinks it has been much over-rated, and that though very large, it is inferior in real value, not only to Sir William's, but to those of Mr. Bentham and Mr. Barker Webb, and perhaps some others. Dr. Hooker believes that the "Vestiges of Creation" were written by Robert Chambers, one of the authors of the "Edinburgh Journal," and "Miscellany." This would explain the otherwise unaccountable secrecy preserved by the author of that work, for the doctrines in it being commonly supposed to be adverse to religion, would give great offence in Scotland, and if Chambers were known to be the author, it might seriously injure or even destroy the sale of his popular works. Mr. Horner observed, in support of this opinion, that there are several Scotticisms in the style of the "Vestiges." Mr. Hewett Watson, the botanist, who is one of those to whom the book has been ascribed, is certainly not the author.

In the evening there was a very pleasant party.

the Miss Moores, the John Phillimores, the Lyells, 1846. Edward Forbes, and many others.

I had much talk with Miss Julia Moore, who is remarkably agreeable. She was a particular friend of poor Blanco White, of whom she speaks with much regard and affection, though, she says that his morbid melancholy and sensitiveness went to the very verge of insanity. When he came to visit her family as he did very often while he lived in London, they were always prepared for twenty minutes of gloom and complaint, and pathetic declamations on his own misery, after which he would gradually brighten up and become extremely agreeable, and even lively. He was a very proud, as well as a shy and sensitive man, and suffered much pain in society from the fear of incurring ridicule, before he had thoroughly mastered the English language. He believed in apparitions.

———

December 5th.

We left our dear kind friends in Bedford Place, and returned to Mildenhall, starting at half-past eleven by the mail train, and arriving at Ely soon after two. Although this railway has got a very bad name, I must say that as far as we are concerned we have never had the least reason to complain of it, and that we find the porters and other people employed on it very civil and obliging.

Our week in London has been delightful, and has done us both good, enlivening and polishing up our minds. Weather very cold and frosty ever since the 29th of last month.

LETTERS.

My Dear Mrs. Horner,

1847   At length I am able to write to you myself, and to thank you, but I can never sufficiently express my gratitude for your kindness (and Susan's too) in coming to stay with us when I was ill.* I was not able while you were here to express at all what I felt, nor indeed to enjoy your society as I should have done at any other time; but I did, and do, and always shall most deeply feel your very great kindness in coming so far at such a season, to cheer and comfort our solitude. I really think my poor dear Fanny's strength would have broken down and she would have been quite unable to go through that time of trial, but for the consolation and support which your company afforded her. My thanks are due to dear Susan as well, and I hope you will tell her that she has made me love her better than ever if possible.

I am still a close prisoner, and the dreary weather gives me no hope of a release. I am quite free from

---

* In the winter of 1846-47, Charles Bunbury had the measles, and was very ill for some weeks —During his illness, Mrs. Horner and Susan came to Mildenhall to help his wife to nurse him.

illness, but do not gain strength, or at very best 1847
slowly indeed, and cannot expect to improve so long
as I continue to be a hot-house plant. It is wearisome enough but I try to be patient for I feel that I
should be ungrateful to my dear wife if I were to
commit any imprudence, after she has taken so
much pains, and nursed me with so much care and
tenderness. Her own spirits are much depressed,
which I cannot wonder at, while she leads such a
lonely and monotonous life. Your letters, and those
of our sisters are among our very best amusements and comforts, and I feel very grateful to the
Horner habit of regular and frequent correspondence.

I have long owed Mr. Horner a letter, in return
for an interesting account of Sedgwick's paper on
N. Wales, which he wrote me a month ago ; but I
have neither spirits nor matter for a letter to him
at present.

If you should chance to see Dr. Hooker, pray tell
him that the Ferns he sent me have contributed
much to cheer my imprisonment.

Pray give my best love to my dear sisters, and
believe me,

Ever your very affectionate son-in-law,

C. J. F. BUNBURY.

<hr />

Barton,
January 26th, 1847.

My Dear Mr. Horner,

I am indebted to you for two very interesting letters, one of which reached me just at the
beginning of my illness, and the other since we

1847 removed hither. I am perfectly well aware what a load of business you have to bear, especially when you are preparing your address, and can thoroughly understand that it must be terribly difficult for you to find time for anything. I am very glad that affairs have gone on so much to your satisfaction, both in the G. S.* and R. S.† in the latter especially I congratulate you on your hard won success. You seem to have got a capital set of new members for the Council of the G. S.‡ I am much obliged to the Council for nominating me Foreign Secretary though it seems rather an odd appointment, considering that I am so little of a geologist strictly speaking, and have no personal acquaintance with any of the foreign bigwigs. However, I will do the best I can.

I have had an interesting letter from Mr. Brown, of Cape Breton, who tells me that he has found " undoubted Fucoids " in several beds of the coal measures there, and promises to send me some.

He says he has found several more fossil trees with Stigmaria roots, and he offers to send me specimens to show the markings, but cabinet specimens, after all will not prove much in such a case, when everything depends upon determining the connection of the supposed roots with the stem.

We have been here a week, and I find myself much the better for the change, and I hope it has done dear Fanny some good too.

I do not know exactly when we shall leave Suffolk, but I hope, at any rate, to see you on our

* Geological Society.   † Royal Society.   ‡ Geological Society.

way to Torquay, even if we cannot remain many 1847 days in London.

Pray thank Mrs. Horner for her kind letter to me of the 20th.

Ever your very affectionate son-in-law,

C. J. F. BUNBURY.

JOURNAL.

February 3rd.

This long interruption to my Journal, as well as to my regular studies, has been occasioned by a serious illness, which attacked me on the 16th December, confining me to my bed for nearly a fortnight, and to my room for a much longer time. It was a complaint rather unusual at my time of life, and a very disagreeable and tiresome one,—the measles; and it left me very weak and languid for a considerable time; but happily it has had no permanent ill effects, and I am now quite recovered.

We have been for the last fortnight at Barton, and I have been much benefited by the change of air and scene. The weather is far from genial, though there have been a few very fine days since we came hither.

My reading has of course been rather desultory and has consisted in great part of novels and other light literature. In fact my illness has defeated all my plans of occupation and study for the winter. I have

1847. however finished " Gardner's Travels in Brazil," of which I had read the greater part before I was taken ill. It is an excellent book of travels.

Since we came to Barton, I have read Gleig's account of the campaign at Washington and New Orleans in 1814 and 1815, an interesting little book, written in a remarkably clear, lively and pleasant style, and very instructive.

My worldly circumstances have been much improved by the share of my uncle Fox's inheritance which has fallen to me.

February 5th.

Began to read the third volume of Arnold's " History of Rome."

LETTER.

Barton,
February 5th.

My Dear Mrs. Horner,

I thank you and my dear sisters most heartily for the pretty presents you have sent me, and still more for your very kind affectionate letters, and for (what I prize more than any gifts) the expressions of your regard and good will towards me. I trust you know that I cordially return your affection

for me, and think myself most happy in having 1847
gained a mother and sisters whom I love and value
so much, and who entertain such feelings of kindness
towards me. I have not time to write separately to
each of the dear girls, but I beg you to give to each
and all of them my hearty thanks. We received the
parcel of presents this morning before we were up,
and had much amusement in unpacking it and
examining the various articles. I hope Mr. Horner's
health is not suffering from his hard work, as I know
there must be a great strain upon him just now,
when he is winding up the affairs of his Presidentship,
and preparing for the Anniversary. I am sure he
will leave behind him the character of one of the
most useful and valuable Presidents that the Geological Society has ever had.

I am delighted to hear of our having Susan's
company in our foreign tour, if we are able to effect
it, of which I flatter myself there is every probability ;
I am sure that Italy will charm her, and it will be a
great advantage as well as pleasure to me to have
the benefit of her opinion on the pictures.

Have you read " Lucretia ? " It is one of the
most powerful, and of the most disagreeable novels,
I ever met with ; the first volume indeed is excellent,
and so are some things in the second, but in the
third, horrors are accumulated upon horrors till one
is quite bewildered. Lucretia, the heroine, is the
most perfectly fiendish character in any work of
fiction that I know of. Altogether, the book leaves
a very unpleasant impression on one's mind, and
indeed has much similarity to the new school of

1847. *French* novels, with this important difference however, that it is quite *proper*.

Post-time draws near, so I must conclude, with my affectionate love to all your family.   Fanny sends her love.

Ever your very affectionate son-in-law,

C. J. F. BUNBURY.

JOURNAL.

February 18th.

We came up to London by the Eastern Counties' Railway, and made out our journey very prosperously. We arrived at the Lyell's house in Harley street, towards dusk, and were most kindly greeted.

February 19th.

The Anniversary of the Geological Society, but I was not able to attend.

Mrs. Horner, Susan, and Katharine, and Edward, dined here, and we had a very pleasant evening. The conversation never flagged.   They gave us some curious and diverting accounts of the very odd and bold conduct of Miss Lister in eloping with Mr. Drummond; how she walked all alone from her stepfather's house in Chesham Place to that of Mr. Drummond's father, in Stratton street, and asked for Mr. Drummond, and finding he was gone out, walked up and down the street waiting for him ; and how, when he came back, she insisted on his going 1847. off with her immediately, in such a hurry that the banks being all closed, he was actually obliged to pawn his watch to provide money enough for their flight. The lady is only eighteen.

Lyell reports that the anniversary dinner of the Geological Society went off exceedingly well, and that there were some very good speeches, especially from Lord Morpeth, Mr. Bancroft, and Bishop Stanley.

Dr. Hooker came to call on me, but could only stay a short time

I had also a visit from my friend John Moore, who was as usual very pleasant.

Dr. Hooker has undertaken to describe the plants collected in the unfortunate Niger expedition by Dr. Vögel. He told me that he had lately obtained a specimen of Lepidestrobus with the internal structure beautifully preserved, and is about to describe and draw it. The vessels are scalariform as in Lepidodendron Harcourtii, and so many other fossil plants.

Babbage called here and talked a great deal. He is full of a theory of volcanic movements. His intellect seems to me remarkably acute and subtle. I think that if he had lived in the middle ages he would have been a great school man.

P

February 25th.

We travelled down to Exeter by the Great
Western railway, leaving Paddington at a quarter
past ten, and arriving at Exeter at about half-past
five ; a very easy and pleasant journey, without any
fatigue. But in such a mode of travelling one can
make no observations of any importance on the
country, except that in the cuttings one can see, in
a large and general way, the succession of formations
—first, the tertiary deposits, extending some distance
from London ; then for a long way, the chalk; next,
the oolite, reaching to a little beyond Bath ; next,
the sandstone and a bit of the coal measures between
Bath and Bristol ; beyond this latter, the extensive
alluvial and marshy plain (singularly level) in which
Bridgewater is situated ; and lastly new red sandstone, all the way from Taunton to Exeter. The
environs of this latter city appear to be very pretty
and extremely fertile.

February 26th.

From Exeter, by railway to Newton, the farthest
part to which that line is still open. The railway
descends along the margin of the Exe to its mouth,
passing close by Powderham Castle, through the village of Starcross, and between Dawlish Warren and
the higher land ; then turning westwards runs along
the very beach, between Dawlish and the sea, and
on to Teignmouth, through deep cuttings and
numerous turnings in the red sandstone cliffs ; and
finally ascends the course of the Teign to Newton.
Here we got into a stage coach and went on to

Torquay, through pleasant lanes, where the Hazel 1847.
was in full flower, and abundance of Ferns, looking
even now quite green and vigorous. Arrived at
Torquay about two, having left Exeter at
twelve.

A very cold day, with a violent N.E. wind, and
no sun, so I cannot stir out. Fortunately the hotel
we are in (Webb's) is extremely comfortable, and
very pleasantly situated. Our windows look
directly on the pier and the little harbour, which
is confined by two jetties built of stone, and beyond
we look across the bay to Paington and Goodrington
sands ; the sea is rough to-day ; the farther shore of
the bay is distinctly marked by a broad line of white
foam, and on some high bold headlands opposite,
the sea is breaking superbly, and rising in sheets of
spray.

Quitted the hotel, and established ourselves in
lodgings, in " Rock House," half way up one of
the hills above the town. Weather still intensely
cold.

A much milder day. The wind having abated
though still in the same quarter. We walked out,
along a very pleasant sheltered path, which runs
along the side of the steep hill on which " Rock

1847. House stands, screened by trees and thickets, and
overlooking the fine extense of the bay, even to
Berry head, and the open channel ; then descending
to the road, and crossing it opposite the Tor Abbey,
we strolled along the sands, the tide being out.   I
had not had so good a walk since the middle of
December, and though the day was not very bright
or favourable, I enjoyed it much.   The form of the
bay is beautiful.

The hills around it are tame in outline, but they
for the most part terminate boldly towards the sea,
forming a succession of bluff headlands, so that the
shores of the bay are indented and cut into several
subordinate bays and coves.   The little town of
Torquay lies in one of these indentations, the
principal part of it down on the shore, beneath and
between the steep fronts of some of the hills, which
form the northern boundary of the bay ; and these
hills, which are partly cultivated, partly planted, and
in some places rocky, are studded all over with
villas and lodging houses.   The general effect is
very gay and pleasing.   Berry Head, which terminates the bay on the south, is a bold and conspicuous promontory, with a rocky escarpment, which as
far as one can judge at this distance, seems to be of
considerable height.   The farthest point that we
could see to the north east, beyond Torquay, is
marked by a very peculiar and conspicuous rock,
perforated by the waves so as to form a lofty arch.

Tor Abbey, a large modern house, belonging to
the Carew family.   A rookery there.

I began to read Henry Napier's " History of

Florence," beginning with the 6th chapter, as I had 1847
read the first five just before my illness.

———

A bright and beautiful day. We walked out in
the same direction as yesterday, but farther, to a
place called Livermead, and then descended into a
charming little cove, screened from the north by a
precipitous rocky headland, and so exposed to the
sun that it was as warm as a conservatory, Here
we lounged for some time, sunning ourselves, and
picking up shells in the pools of clear water left
among the rocks by the retiring tide. The cliffs in
this part are of dark red conglomerate, like that at
Dawlish and Teignmouth, with some strata of fine-
grained variegated sandstone, the stratification
remarkably distinct and very striking to the eye.
The hill on which our lodging-house stands, is of
limestone, and a fine escarpment of this rock is
displayed immediately above the main road, near
the turnpike, in the way from Torquay to Tor
Abbey.

I was delighted to-day with the scenery, which the
bright clear sunshine brought out to the best
advantage. The innumerable white houses so
thickly scattered over the sides and tops of the green
hills, had quite an Italian brightness and cheerfulness
of aspect ; the beautiful bay was smooth and blue,
its waters rolled gently in on the sands with that
quiet splashing sound which is so soothing to the ear
and the imagination ; and in the latter part of the

1847. day, the effects of light and shade on the fine rocky capes which bounded our view to the north and south, were very beautiful.

Read a very interesting article in the *Edinburgh Review*, on the genius and writings of Pascal,— written, I understand, by a Mr. Rogers, a clergyman at Birmingham.

A beautiful day. I strolled out along the beach, and round the first headland to the south, into the cove near Livermead, and spent some time in hunting for sea-weeds. The rocks of this headland have been worn and cut by the waves into caverns and hollows of a most curious and romantic aspect ; they look like fit retreats for mermaids and all kinds of sea-monsters. In one place a complete archway has been formed. The diversified colours of the sandstones and conglomerates have likewise a very singular effect.

We made an excursion (the weather being very bright and fine) to Babbicombe, two miles off, Fanny being conveyed in a little donkey carriage, while I walked. A long ascent brought us on to the high grounds on the north side of Torquay, from whence gradually descending, through pleasant winding sylvan lanes, we came suddenly full upon a most beautiful little secluded cove, almost encircled by bold cliffs of grey marble, interspersed with smooth steep grassy slopes; rocks of singularly wild and striking forms, jutting out into the sea at the entrance of the basin. The sea was just now of the brightest blue, and the effect of the whole scene, coming suddenly on it as we did, was almost magical. The romantic spot is called Anstis Cove. The Bishop of Exeter's residence, Bishopstowe,—a handsome house built in the style of an Italian villa, is delightfully situated on the hill above.—Proceeding about half-a-mile farther, we arrived on the downs directly above Babbicombe Bay, from whence we had a fine view of the whole line of coast stretching away to the north-east, even as far as Lyme in Dorsetshire,—that town being plainly discernible. The sun shining on the distant red cliffs near Sidmouth, gave them a strangely vivid appearance. We descended by an excessively steep road to the sea side; but the tide was in, and we could not ramble far along the beach,—the weather had by this time changed, the wind was very cold, and a hail storm coming on, obliged us to return rather hastily. I think we should have been more struck with Babbicombe, if we had not just before seen Anstis Cove; yet it will be well worth another visit. —I saw Babbicombe in August, 1826, when travelling with my father and mother, and perhaps the impression that had remained in my memory was somewhat exaggerated.

1847.   I observe that beautiful Fern the Hart's-tongue, in
remarkable profusion and luxuriance in the hedges
and hollow ways hereabouts; Aspidium aculeatum
also frequent; Rubra peregrina, Iris fœtidissima,
very plentiful in similar situations. I have not yet
looked much for mosses in this neighbourhood, but I
have found Pterogonium! Smithii (a rarity in
England) on more than one tree, and Neckera
heteromalla seems to be frequent.

_____

March 12th.

This day and yesterday were quite delicious. We
walked out to the headland directly above Meadfoot
sands, on the E. of Torquay, and enjoyed the view
exceedingly. After rambling about the open ground
at the top, which is covered with that fine, short, close
turf, characteristic of a lime-stone soil, and sprinkled
with tufts of furze, now in full blossom,—we
descended to the edge of the cliff, and sat some time
basking in the sun, and enjoying a warmth like that
of summer. The sky was cloudless, the sea of a
Mediterranean blue, and perfectly calm, just curling
and rippling round the rocky islets that lay beneath
us. Above and around us were noble picturesque
rocks of grey marble, veiled here and there by ivy
and various shrubs. The whole scene was delicious,
and my enjoyment of it was doubled by witnessing
the pleasure it gave to the dear partner of
my heart.

_____

I walked to Preston Sands, and along them as far
as Paington.   In the way, looking down into one of
the many little coves with which the coast is indented, I saw a kingfisher,—it looked like a little
flash of bright blue light, darting from one rock
to another, and it was not until perched that I
distinctly made out what it was.

Weather lovely.   I walked again to Anstis Cove,
and rambled about it for a considerable time, and
then walked along the fine open level down above it
till I looked down on Babbicombe.   The sky was
cloudless, but there was wind enough to produce a
strong fringe of foam along the rocky shore.   The
limestone rocks around Anstis Cove are magnificent,
in some places rising in wall-like precipices, in
others standing out like great towers; and between
them are slopes covered with short, close turf, intermixed with thickets of furze and brambles.

I observed on the downs above this cove, some
singular clefts or chasms in the lime-stone rocks,
some on a small scale, others apparently penetrating
to a considerable depth and extending many yards
in length,—all evidently produced by some natural
cause.   I had noticed similar figures in the limestone
of the headland above Meadfoot Sands.

I was struck with the scarcity of cryptogamic
vegetation on these limestone rocks.   I saw very
few mosses and none in any abundance except the

1847. two beautiful kinds—Neckera Crispa and Hypnum Molluscum—which are characteristic of calcareous soils, and which here, as well as about Genoa, grow in thick tufts among the grass on the more shady parts of the hills and around the smaller rocks that rise above the soil. This reminded me of the delight with which I gathered Neckera Crispa for the first time, more than 20 years ago, in this very neighbourhood. In the chasms above-mentioned, I observed a few other mosses and Jungermanniæ; but the great towering rocks seemed entirely destitute of cryptogamic plants, and scarcely even discoloured by a lichen. In flowering plants, however, I have no doubt this spot would be very rich at a favourable season, and would afford many varieties.

March 16th.

A stormy day with rain and a high wind setting directly in shore, so that the sea rolled in handsomely. The tide rose remarkably high, and the harbour was filled almost up to the level of the jetties. Before the rain set in, I walked down to the Abbey Sands and examined the curious submarine forest, of which the remains are very visible at low water: numerous stumps of trees and many of very considerable size, are seen standing up out of the sand, but so covered with sea-weed that they may easily be taken for rocks. The stumps are accompanied by large accumulations of peat, or something very closely resembling it. In some places there are great clusters of small stumps, like the remains

of thickets or copsewood, mixed up with clay and 1847
peat.

———————

Again to the Abbey sands, and the rocks immediately beyond them. The flat shelving rocks below high-water mark, are everywhere covered with a most profuse growth of Fucus vesiculosus and serratus, Fucus nodosus occurs more partially. Halidrys siliquosa, Laminaria digitata, and Laminaria saccharina, are thrown up on the sands in great quantities, and the last of these three I have also seen growing plentifully in the pools left among the rocks at low tide. Laminaria bulbosa is frequently thrown up on Preston Sands, and Himanthalia lorea on those of Tor Abbey. Rhodomenia palmata is very common. On the rocks not far above low-water mark I have already collected several of the smaller and more delicate kinds, in particular a beautiful little Chylocladia (articulata, I think), looking like a miniature Opuntia, but of a delicate transparent pink colour. I observe here that the sea-sand, being very much mixed with fragments of shells and zoophytes, and with sea-weed, is in great request as a manure, like the Bantry Bay sand in Ireland.

I met two boys carrying a large basket full of a species of Solen, the broken shells of which are very frequent on the beach.

———————

March 18th.

Another delicious day. We went to Paington
Sands, where we saw a number of persons, principally
women, collecting the Solens, which they call (as we
understand) Hut-fish or Spear-fish, and which are
used for food. These shell-fish are very numerous
here. They live in the wet sand, where it is unmixed with rocks or shingle, and not very far above
low-water mark, burrowing vertically in to it to some
depth. Every now and then one sees them rise up
perpendicularly with a sort of spring, just showing
themselves a little above the surface, and disappearing again; but the fisher watches the spot
where they jumped, (which is marked by a sort of
dimple in the wet sand), and instantly digging down
with a broad wooden scoop, brings the shell-fish to
light, and pulls it up with his hand. I watched a
woman taking many in this way, and I observed that
she had to exert considerable force to pluck the
creature up, so that its muscular power must be
very great in proportion to its size. Sometimes also
they spear the Solen with a pointed steel rod, which
they thrust down through the sand. We found also
some fine large Cardiums with the animal in them.
Some of the low flat shelves of rock, exposed at
low water, are entirely covered with mussel shells, as
close as they can stick, so that one cannot put in a
little finger between them; they adhere firmly to the
rocks, in a vertical position, with the hinge downwards. We observed likewise great numbers of white
shells, (a species of Pholas, I believe), lodged in
holes, in some of the sandstone rocks.

Read John Phillimore's pamphlet on the Reform
of the Law; good, but (as it was to be expected
from him) rather too vehement.

March 22nd.

We called upon Mrs. Griffith, so famous for her
profound knowledge of Algæ, and skill in preserving
them, and so often mentioned in Dawson Turner's
Historia Fucorim, Harvey's Manual, and other
works on that subject. Sir William Hooker had
given me a letter of introduction to her. We found
her a very intelligent and spirited old lady, with a
frank, pleasant manner, and still full of eagerness
about her favourite study.—

Lady Bunbury and Cecilia arrived here in the
evening.

March 27th.

Visited Mrs. Griffith again, and spent two hours
in looking over a part of her magnificent collection
of the Algæ of this coast. The specimens are most
beautifully prepared, and peculiarly instructive,
generally exhibiting complete series of all the
varieties and stages of growth of each species.
Many of them are exquisitely beautiful in form and
colour.

April 3rd.

We went to an early dinner or luncheon party at
the Buller's, and met Charles Buller, who had come

1847. down from London to stay with his father and mother during the Easter recess. He was very entertaining and agreeable.

LETTER.

Rock House, Torquay,
April 6th, 1847.

My dear Leonora,

I am much obliged to you for your letter, and very glad to hear that my remarks on the Pines were of use to you. I rejoice to find that you are studying botany so zealously and diligently, and in such a judicious way, taking up one family at a time and thoroughly working it out. The opportunity of conversing with Mr. Brown, and consulting the British Museum Collections must be very valuable helps to you. The Ferns on which you are now at work are my especial favourites, and when I come to town I shall be very glad to discuss them with you. I suppose you will go to Kew and examine the beautiful collection there.

I have been studying the seaweeds a little during our stay here, and have collected some, and have improved in the art of preserving them, thanks to Mrs. Griffith's hints. This is not the best season for them, as most of the delicate kinds are in their greatest beauty in summer, and many are only to be found then, but even now, besides the large coarse kinds, ·such as the Fuci and Laminiariæ, which

abound at all seasons, I have picked up a few very 1847
delicate and beautiful little species. I think I have
got about twenty in all, besides a few beautiful
Sertularias—and of most of these I have plenty
of specimens to spare. But now, alas, my seaweed
hunting is at an end, for I have had a little attack
of illness—inflammation in my face and eyes, with
some degree of fever,—and the doctor pronounced
that I must not get my feet wet any more, nor
stand dabbling in the pools of salt water with my
head stooped ; and unfortunately the best Algæ are
only to be found among the rocks at very low water,
and not to be attained without wetting one's feet.

The Hart's-tongue Fern grows in greater profusion
here, and more luxuriantly, than I have ever seen it
before, and is very ornamental. I have seen
asplenium marinum in two or three places, but
always out of my reach.

I have been reading Meyen's " Geography of
Plants " (one of the volumes published by the Ray
Society) with great satisfaction: it is full of interesting
matter, much of it derived from his own observations
in his voyage round the world; though I do not
always agree with his views, and he has been led
by Lichtenstein into some considerable errors respecting the vegetation of the Cape. I strongly
advise you to read the book, as you can borrow it
from Charles Lyell.

I am growing rather desirous to be in London,
but must not attempt to go there till I hear that
you have mild and settled weather. During
part of the time Lady Bunbury was with us, the

1847. weather was delicious, and so was last Wednesday,
but on Saturday we had snow and winds worthy of
December.   However, I do hope we shall get
to London in time to see Katharine before she
starts for Switzerland.   Love to all.

Ever your affectionate brother,
C. J. F. BUNBURY.

JOURNAL.

April 18th.

I walked to " Hope's Nose," and thence along the
coast to Meadfoot Sands, and so home.   Examined
the interesting geological phenomenon of the raised
beach near "Hope's Nose," well described by Mr.
Austen in the Geological Transactions.   It is
situated about a quarter of a mile or so to the south-
west of the point properly called "Hope's Nose,"
almost directly in a line between it and the remarkable island rock called the "Thatcher." It forms here
the top of the low cliff, resting on the limestone at
about 30ft. above the sea ; is composed of much the
same materials as the present beaches in Tor Bay,
but consolidated into a hard stone.   I was much
interested by the examination of this, the first
example I have ever seen of the kind, and a very
well-marked one.   The resemblance to a recent sea-
beach is so perfect, that one feels it is in a manner
an occular demonstration of the change of level.

The limestone rocks within reach of the tide, on

this part of the coast, are curiously honey-combed 1847. by the action of the water. At "Hope's Nose" there is an extensive quarry of beautiful black and variegated limestone like that of Babbicombe, and indeed the promontory has been cut down to a great extent by the quarrying.

The rocky islets which lie off this part of the coast add much to the picturesque effect of the scenery, especially the one called the "Thatcher" Rock, which is of considerable height and crowned with fine castellated crags of a very striking aspect.

April 20th.

This is to be our last day at Torquay, and though we expect much enjoyment in London, both Fanny and I feel considerable regret at leaving this pretty place. To day, too, it seemed to put on its very best looks, as if on purpose to make the most favourable impression on us just before our departure; for the day was extremely fine, and the bay looked more bright and blue, and the whole range of coast was seen with more vivid clearness than has been the case for nearly a month past.

April 21st.

Came up to London by the express train in six hours, from Newton to Paddington. Arrived safe and sound, and settled ourselves in lodgings.

April 22nd.

Walked about the town and saw sights ; looked at

Q

1847. the British Institution,—the worst exhibition I remember ever to have seen there. Went to the
Geological Society, and had a pleasant chat with
John Moore. We dined with the Horners, and
enjoyed a most comfortable and merry talk with the
girls. Poor Mr. Horner far from well. Went to
an evening party at Mr. Hallam's.

I hear a very satisfactory account of my paper on
the Richmond Fossil Plants, which was read on the
14th, together with Lyell's geological account of
that coal field. I hear that the discussion was
excellent, and that my paper was highly approved by
those most competent to judge of it.

April 23rd.

Henry Napier called on us very soon after breakfast; he is in bad health. He was scarcely gone,
before Clarke and Sir Anthony Oliphant made their
appearance ; the latter I had not seen since he left
the Cape. Clarke gave me a tolerably good account
of dear Sarah. He told me a good deal about the
unfortunate Cape colony and the mismanagement
of the Caffer war, and especially the extraordinary
conduct of Stockenstrom. That man having been
entrusted with the command of the Burgher force
employed in the invasion of Cafferland, instead of
co-operating with the other divisions of the army,
as it had been intended that he should, chose to
conclude a treaty on his own authority with Creili,
the son of Hintza ; by which treaty Creili was
acknowledged as the paramount chief and sovereign
of the Amakosa Caffers, and in return promised to

yield up to the English all the territory lying to the 1847
west of the Kei river. Now, what is extraordinary
in this is, that these very same conditions formed
a principal part of the treaty concluded by Sir
Benjamin D'Urban with Hintza (Creili's father), ten
years ago ; and that Stockenstrom on that occasion
not only inveighed vehemently against the treaty,
but induced Lord Glenelg to annul it ; urging (and so
far with great truth, I believe), that Hintza could in
no sense be considered as paramount Chief of the
Caffers, or as having any authority over the other
chiefs,—that he had no right to cede the territory
west of the Kei,—and that his cession of it was
a mere farce, without any sort of validity. Sir
Peregrine Maitland disavowed the treaty thus concluded by Stockenstrom, upon which the latter threw
up his command and assumed the tone of an injured
patriot.

After luncheon I went out, paid many visits, and
saw the Suffolk street exhibition of pictures, which,
though not good, is better, this year, than the
British Institution. "The Close of a Selfish Life" by
Prentis (the idea taken from a scene in " The King's
Own,") is a striking and forcible picture.

We went in the evening to a delightful party at
the Lyell's. Agreeable as their parties almost always
are, I do not remember any more agreeable than this.

I was introduced to Dr. Wallich, the famous
Indian botanist, whom I found a very interesting
man, with a fine countenance and pleasing manners ;
we talked much about the Cape, where he had spent
a considerable time, a few years after my visit, and

1847. had travelled all over the Colony, and collected
extensively.   He spoke with enthusiasm of Sir
George Napier.   It appears from what he told me
that the fossil plants of the Burdwan coal field, are
still but very partially known, and that much yet
remains to be investigated.

I had much talk also with John Phillimore on
reform of the law, a subject on which he feels very
strongly, and, I think, very justly.   He is very
zealous for codification (as Bentham calls it), and is
a great admirer of the Roman law, though he admits
that it was very natural our ancestors should oppose
it, since those who wished to introduce it into this
country, favoured it, not on account of its systematic
merits, but because it was supposed to be more
favourable to arbitrary power.   He told me that the
study of the Roman law was pronounced by Leibnitz
to be equal to that of mathematics, as an exercise
and discipline of the mind.

April 24th.

Saw the panorama of Cairo, which is interesting,
and gives one a lively idea of the peculiarities of an
oriental city.

We all went to Babbage's evening party, and
enjoyed it much.   There was a great variety of
people.   I met many whom I knew and liked, and
had much pleasant talk, especially with Sir Edward
Ryan, John Moore and his brother, Dr. Falconer,
and Mrs. Milner Gibson.   The latter is intimate
with Disraeli, and I talked with her about his
strange and fantastic but very entertaining novel of

"Tancred." She says Monckton Milnes is affronted 1847. at the portrait drawn of him (under the name of Vavasour) in the first volume of that book, which she is surprised at, but I am not, for though not an unfavorable picture on the whole, it is drawn in a quizzing style. The character of Fakredeen, she tells me, is drawn from Mr. Smithe, one of the young England party.

Dr. Falconer told me that Joseph Hooker has made great discoveries respecting the fossil cones called Lepidostrobi, completely confirming Adolphe Brongniart's notion, that they were the fructification of Lycopodiums or plants very nearly allied to the Lycopodium.

April 25th, Sunday.

Went to afternoon Church; afterwards went to call on Louisa Napier, and saw at her house a good portrait of her brother, Sir Charles.

In my walk afterwards, I met Edward, who in the course of conversation mentioned that Tom Young ("Ubiquity Young") had told him he knew *positively* that the "Vestiges of Creation" were written by Robert Chambers of Edinburgh. It seems very probable.

We went to a quiet tea party at the Horner's, where I was introduced to Mr. Erskine, a very clever and highly accomplished man, and in particular a distinguished oriental scholar. He is very pleasing, of great modesty and simplicity of manners, and very communicative. We had much talk about India, and among other things of the great railroad

1847. which is designed to be carried from Calcutta to Delhi. Mr. Erskine thinks that the principal impediments in the way of it (and very serious ones) will be occasioned by the numerous rivers, tributaries of the Ganges, which in the rainy season, become flooded to a great extent, some of them spreading over the plains to the breadth of three miles—and which then come down with great fury.

Mrs. Erskine, a daughter of Sir James Macintosh, is a very agreeable person.

April 27th.

I paid sundry visits; among others to Mrs. Romilly and Lady Charleville; with the latter I was scarcely acquainted before. She is a very talkative and entertaining old lady, of extensive reading as well as great knowledge of the world, clever and witty.

April 28th.

We breakfasted with Mr. Rogers, who was extremely courteous to us, and made himself very agreeable, as indeed I have always found him on such occasions. We three were alone with him ;* Lord Glenelg had promised to come, but did not.

Afterwards I went to the Council of the Geological Society, where a rather active debate took place, on some propositions of Daniel Sharpe; and in the evening to the general meeting of the Society. A paper by Captain Vicary, on the Geology of Scinde, was read.

* Charles Bunbury and his Wife and Cecilia Napier.

It was dry enough in itself, being little more than a 1847
collection of rough notes, but it gave occasion to a
very interesting lecture from Dr. Falconer, on the
Geology and Physical Geography of the northern
and north-western boundaries of India. He pointed
out the striking agreement in geological structure
between the Hala and Sulimaun mountains, which
run northwards from near the mouth of the Indus to
within sight of Caubul, and the Siwalik mountains,
which form the outer or advanced range of the
Himalayas. He noticed also the prodigious development of the nummulite limestone and of the *bone
beds* or very late tertiary deposits, in which the remains of large extinct mammalia were found in such
extraordinary abundance and variety by himself and
Major Cautley. These bone beds have been traced
at intervals along the Hala range and the other mountain chains connected with it, for nearly a thousand
miles from south to north, and along the Siwalik
hills for 1700 miles from north-west to south-east.
He then remarked the great contrast presented by
the actual condition of these mountain-chains,
which appear from geological evidence to have been
formerly so very similar ; the Siwalik hills being
covered with luxuriant forests, while the mountains
of corresponding geological structure on the West of
the Indus, are utterly bare and barren, like skeletons
of mountains, without a tree or a shrub, or a green
spot, and almost without a drop of water. He gave
us a striking picture of the utter desolation of those
mountain tracts, and of the wild and fierce character
of their human inhabitants. Finally he noticed the

1847. great salt range cut through by the Indus at Kalabaugh, which he considers of secondary age; and he mentioned the existence of Palæozoic rocks in the mountains east of Caubul. Murchison observed that the remarkable conformity of geological structure between the Sewalik hills and the mountains west of the Indus, while their directions are so entirely different, is adverse to Elie de Beaumont's theory of the parallelism of contemporaneous mountain chains.

April 30th.

Went in the evening to the Royal Institution, and heard an excellent lecture from Charles Lyell, on the Age of the Volcanoes of Auvergne.

Lyell delivered this lecture admirably well, and had a very attentive audience.

May 5th.

Went to the Exhibition of the Royal Academy. Etty has a magnificent work, a triple picture, or series of three pictures in one frame : the subjects taken from the life of Joan of Arc. In the first, she is represented as finding the sword which had been indicated to her by a vision, and devoting herself to the service of her country ; in the second picture she is sallying from the gate of Orleans, and scattering the English ; and in the third (the finest of all) she is chained to the stake, and raising her eyes to heaven, while the flames are beginning to curl up around her.

These are really grand compositions, especially the 1847. second and third, and give me a higher idea of Etty's powers than anything else I have seen of his. He has another picture in this exhibition,—the Graces,—the flesh well painted in his usual rich and voluptuous style, but the faces vulgar and meretricious. A most beautiful picture by Stanfield, —Sarzana and the Carrara mountains, from the Magra. Landseer has two pictures,—one, a Deer-stalking scene in the Highlands, in which I cannot see any particular merit ; the other, Van Amburgh among his wild beasts,—a strange subject, selected, it is said, by the taste of the Duke of Wellington ; the lions, tigers and leopards are beautifully painted. But the most charming picture in the exhibition, to my mind is one by Frith, a subject from the Spectator : —Sir Roger de Coverley showing the Spectator his portrait, metamorphosed into the Saracen's Head, on a sign-board. Sir Roger is a perfect representation of the honest and simple gentleman that is described by Addison, and the grave earnestness with which he is pressing for his friend's opinion, is very well contrasted with the diverted look, the humorous but quiet laugh of the other. It is altogether a most amusing and pleasing picture ; the story is perfectly well told, without the least exaggeration or affectation ; the two principal personages have completely the air of gentlemen ; the host too is good in his way, and all the accessories are well worked up.—There is also a clever and humorous thing by Webster,—the clerk and choir of a village church, preparing to sing. A

1847. pleasing and rather touching scene of domestic life,
(entitled, "The invention of the Stocking Loom,")
by Elmore. A good portrait of General William
Napier by Mogford. These were all that I
particulary noticed in this first visit to the
Exhibition.

May 6th.

Went to the Linnean Society, and spent some time
in studying Dr. Hooker's Flora Antarctica. In my
last visit there, two days before, I had noticed a
very odd-looking old man, as small as a Bushman,
and singularly ugly ;—Mr. Kippist now gave me this
account of him,—that he has the peculiar monomania
of spending all his time in copying and recopying the
text of Sowerby's English Botany,—and thus he
has gone on for years. He is generally to be seen in
the reading-room of the British Museum, employed
day after day in his voluntary task of copying ; but
the Museum being just now shut, he had recourse to
the Linnean Society. We dined at Mr. Lloyd's,
and went afterwards to a party at the house of
Chevalier Bunson, where we met Sir John and Lady
Herschel, Mr. and Mrs. Erskine, and some
other friends.

May 7th.

We had a pleasant dinner party at Sir Charles
Lemon's ; met Lady de Dunstanville, Lord Ilchester,
Sir Edmund and Lady Head, Mr. and Mrs. Vivian,
Mr. and Mrs. Trelawny.

Dined with the Richard Napiers, to meet my father, who leaves London to-morrow :—he and Lady Bunbury, Edward, Mr. and Mrs. William Grey, and John Phillimore formed the party. Much political talk after dinner,—Grey and Phillimore agreeing that the commercial and financial situation of the country is very alarming.

May 10th.

Visited the water-colour exhibition, where I saw several very pretty things, especially a very fine view of the Isle of Staffa in a storm, by Copley Fielding ; — a view of Snowdon, from the road between Tany Bwlch and Pont Aberglaslyn, by the same ;—" The Stag at Bay," by Frederick Taylor ;—some excellent Highland Scenes, by Evans,—and a Moonlight View on the Coast of Cornwall, by Gastineau. Then I went on to the Royal Academy exhibition, and found Lady Bunbury and Fanny there. We dined with the Lyells ; a family party, with the addition of Mrs. Doormann, a German lady. Lyell talked much to me about an article in the "Westminster Review" on Biblical Criticism, and on the theories of Strauss and other rationalists,—very remarkable, he says. He spoke with great praise of Hugh Miller's new book.

May 11th.

In a second and third visit to the Royal Academy Exhibition, I have been less pleased with Etty's

1847. great picture, at least with the central part of it, than I was at first.   Yet still with all its faults, it gives me a much higher opinion of Etty's powers than those continued repetitions of the same naked figure in various postures with which he has usually contented himself.

Maclise exhibits a large picture,—" Noah offering sacrifice on coming out of the ark," which does not please me at all.

Turner displays the most extravagant and incomprehensible, I think, of all the wild vagaries of which he has grown so fond.

There are also several scriptural attempts in this exhibition, but none of them very successful, I think.

Our English artists shine much more in humorous subjects and those of domestic life.   Webster's "Village Choir" is capital, especially the clerk who is giving out the Psalm, and who is every inch a parish clerk.

The "South Sea Bubble" by Ward, is also an excellent picture, remarkable for variety and force of character, and the appropriateness of expression in all the various faces.

On the other hand Mulready's " Burchell and Sophia Haymaking " (from the Vicar of Wakefield) appears to me very insipid.

Etty has received £2,500 for his triple picture of " Joan of Arc."

We went to the Flower-show of the Royal Botanic Society, in the Regent's Park. It was a beautiful sight. The flowers magnificent, especially the Azaleas and the Orchideous epiphytes ; among these latter, several fine specimens of the Phalænopsis, with its exquisitely white flowers ; some most beautiful Cattleyas ; the Calanthe veratrifolia, and Vanda teres, were conspicuous.

There was also a very beautiful display of Cape Heaths. The chief novelty to me was the Tropæolum azureum, a delicate climber, much like the Tropæolum tricolour in its general aspect, but with violet or bluish purple (rather than azure) blossoms.

The number of people present at the show was very great, and the gay-coloured bonnets and parasols of the ladies in the bright sunshine, had a very pretty effect. I dined at Richard Napier's, and went late to the Geological Society, where, however, I heard a very good discussion, on the subject of the human remains in Kent's Cavern at Torquay. The question was, whether the human bones and works of art found in that cave were contemporaneous with the remains of extinct quadrupeds, with which they occur apparently mixed. Austin maintained that they might be contemporaneous, and Sedgwick appeared somewhat inclined to the same opinion ; but Buckland, Lyell and Mantell took the opposite view, and seemed to have the general opinion of the meeting on their side.

To the British Museum. The new entrance by the grand portico in the centre of the front is now open to the public, though the quaint dingy old gateway to the street still remains standing. The portico is on a grand scale, and a very fine thing in itself, but perhaps it too much eclipses the building to which it belongs. The entrance hall too, is fine, but I do not like the bare blank wall which meets our view on going up the first flight of the staircase. It ought to be relieved either by niches with statues, or by fresco paintings. The arrangement of the antiquities is still in progress, and some of the rooms not yet opened. The collection of Greek (or Italo-Greek, or Etruscan) vases is a very rich and very interesting one, and remarkably well arranged with labels explaining the subjects of the designs, when they can be made out. I called afterwards on Mr. Brown, at his house in Dean Street, and had a long and pleasant talk with him. I found him very agreeable and very communicative, as I have usually found him in a *tête-à-tête*, when he is at his ease, and talks freely, without fear of being overheard. In society he is either shy or reserved, and unwilling to talk. His rooms exhibit an overwhelming profusion of scientific wealth, in the greatest possible disorder, even exceeding what I have witnessed at Mr. Stokes's. In the evening, went to the Royal Institution, and heard Forbes's lecture on the Natural History of the North Atlantic, its zoological provinces, and more particularly its fish and fisheries.

We dined at Bedford Place,—a family party.
Poor Cecilia taken frightfully ill in the night. I
went down in a cab, at 3 o'clock in the morning, to
Chesham Place, to fetch Lady Bunbury.

———

May 20th.

Went to Kew by the omnibus, and spent nearly
three hours by myself in those magnificent gardens,
which were now in full beauty, the recent fine and
warm weather having brought vegetation very
forward.

I lingered long amidst the superb collection of
Ferns. The new Palm-house, on a grand scale is far
advanced towards completion.

———

May 21st.

Dined with Sir John Boileau, went with him to
the Royal Institution, and heard a very good lecture
from Mr. Sidney on the Parasitic Fungi, which
cause dry-rot, and those which attack various kinds
of food and household articles.

———

May 23rd.

We dined at Bedford Place,—met Major and Mrs.
Power, Mr. Brown, Mr. Forbes, Mr. Stokes and Mr.
Babbage. Owen and one or two others came in the
evening. I had much very pleasant talk with

1847 Babbage and Forbes; also with Mrs. Power (Fanny's aunt), who is a clever and accomplished woman.

May 24th.

I went with Mr. and Mrs. Horner, Major and Mrs. Power and Joanna to see Mr. Vernon's fine collection of modern pictures in Pall Mall. There are the finest Turner's I ever saw, a very fine Gainsborough, Eastlake's "Christ mourning over Jerusalem," Callcott's "Returning from Market," Landseer's "Time of Peace" and "Time of War," (the two companion pictures which were in the Exhibition last year), Jones's "Battle of Corunna," and capital pictures by Leslie, Mulready and Etty.

I dined with the Linnean Society, at their anniversary dinner—much speechifying, the Bishop* of Norwich (who was in the chair) spoke very well. I had some good talk during dinner with Mr. Miers (the investigator of Chilian botany), and afterwards with Edward Forbes.

May 25th.

We went to Kew, with Leonora and Joanna, and enjoyed it exceedingly, the day being exquisite and the gardens in their highest beauty. There are noble trees of Quercus Cerris, Juglans nigra and Sophora Japonica; a very fine deciduous Cypress, I should think 6oft. high, very symmetrical and graceful in its form; a Deodara, young, but very flourishing, 12 or 15 feet high; Magnolia grandiflora

* Edward Stanley.

standard. The Abies Douglasii and Magnolia 1847 acuminata not nearly so large as at Barton. The great Araucaria imbricata not in the least injured by the severity of the late winter. In the houses, we particularly noticed the Cocos coronata, a magnificent Palm from Brazil;—the Cecropia peltata, so common in the Brazilian forests,—the Coccoloba pubescens, with immensely large round leaves finely veined with red;—the Nepenthes Rafflesiana, a very fine new Pitcher plant, with Pitchers far larger than those of the old Nepenthes, and beautifully variegated with deep purple;—the magnificent Ferns, Dicksonia squarrosa from New Zealand, and Dicksonia antarctica from Van Diemen's Land;— Bignonia Colei, remarkable for its delicate flowers growing out of the old trunk and thick branches; —Eucalyptus macrocarpa, a new kind from Swan River, showing its large flower-buds with their curious lid-shaped calyx in the axils of the large glaucous leaves;—and several beautiful Grevilleas in full blossom.

[The illness of his cousin Miss Cecilia Napier, the arrival of a little nephew from Australia, and his brother Edward's canvassing to be M.P. for Bury St. Edmund's, interrupted his journal and all his usual occupations. He and his wife were living more at Barton than Mildenhall during the months of June and July.]

R

LETTERS.

———

My Dear Mr. Horner,

1847.   I am very glad that your German tour has
been so agreeable and so interesting as it seems to
have been, and should hope that the amusement and
change of scene, and respite from hard work, must
have done good to your health.   I am sorry you are
obliged to be in the manufacturing towns during this
boiling heat, for though I think London in hot weather
is about as pleasant as purgatory, yet I suppose
Manchester must be many degrees worse.   I am
exceedingly sorry that you cannot give us any hope
of seeing you here before we go abroad, for it would
be a very great pleasure to both of us, and I think
you would enjoy this place, which is now in its best
looks, and as cool and fresh as any place can be in
such weather. But you are undoubtedly the best judge
of what you can do consistently with your official and
other engagements. Our garden is very successful this
year ; we have flowers in profusion, and (what we
have not had before during our residence here)
abundance of fruit.

The school is very flourishing.   We have been so
much at Barton this summer in consequence of

Cecilia's illness, that it has a good deal interfered 1847 with our regular occupations. Fanny, in particular has had little time for study.

I have been mostly engaged with my Cape Notes, to which I have written an introductory chapter, and I am anxious to get the whole finished before we start from hence.

I have done little in the way of fossil botany. In recent botany, I have been engaged with the Ferns and the Fuci.

I have lately written a letter to the Bury paper on the subject of the Poor Law, and will send you a copy of it as soon as it is published.

A little nephew of mine, my brother Hanmer's son, has been staying with us now some time, and is to remain under Fanny's charge till he goes to school. He is a very fine little boy, very strong and active, and full of animal spirits, and at the same time very frank, open and good tempered, and uncommonly docile for a child of his age.

Edward's canvass has been most successful; the Chartist opposition which was at one time threatened has vanished.

With respect to the question of my coming into Parliament, it is one that may *keep*, for there is certainly no chance at this election. Ten years ago I was certainly very eager to come into Parliament, and made great exertions for that object; but it was a time of strong political excitement, when great and spirit-stirring questions were in agitation, and when the House of Commons had something to do beyond voting money for the Irish, and deciding on the

1847. merits of railways. I can conceive nothing more wearisome or uninteresting than such a session as this last has been. I have no ambition to be a member of a Railway Board. This state of things may change in the course of the next few years, questions of more stirring interest may arise, and my political ardour may revive, but at present that field seems to me " stale, flat and unprofitable."

I do not deny that there are questions of real interest to be settled, but at present they seem to have no chance of a hearing between the clamour of Irish distress and the struggle of rival railways.

Believe me,
Your very affectionate son-in-law,
C. J. F. BUNBURY.

From his Father.

July 1847.

My Dear Charles,

Your letter to the *Bury Post,* pleases me *thoroughly :* and I am very glad that you have written it, and very glad that you make the public acquainted with your opinions. The local resolution of the guardians at Mildenhall is, as you say, an *expost facto* regulation ; and so in truth was the New Poor Law, which said in 1834 to some hundreds of thousands of old people, " you shall be constrained for the rest of your lives to exist in a workhouse, or starve, unless you have in the days of your strength laid by the means of supporting yourselves in old age ; though you have never been forewarned of the

coming of the new Law, and the practice under the 1847.
old Law forbade your saving money."

I am inclined to believe that Twiss will not come
to the scratch. According to Leech's dotting-up,
Edward would beat the interlopers by above a
hundred.

"Brother Ned" is a capital canvasser, and has
won golden opinions in Bury. Much love to Fanny.

Ever affectionately yours,
H. E. B.

Barton,
    July 13th.

From Lady Bunbury.

July, 1847.

My Dear Charles,

Your father has just let me read your
answer to "Agricola," and I like it so much that I
cannot refrain from telling you so, though I am half
dead with the heat! In addition to the manifest
hardship to the feelings of the poor which you so
humanely enter into and describe, there always
seems to me one unanswerable reply to the assertion
of its being *no* hardship ; if it is not a hardship how
is it *a test?* If they are so much better off in the
Union House than in their own cottages, why do
the guardians hold out such a reward for the sin of
poverty ?

There are no people who blow hot and cold "like
the advocates of the perfectibility of the New
Poor Law !"

1847.　My kind love to Fanny, and thanks to her for her affectionate letter on my birthday ; I shall not forget hers on the 18th.

My precious Cissy was certainly better yesterday, but still so weak and languid that Dr. Hake does not yet speak of moving her.—I had a good account this morning.

Adieu, dear Charles, and believe me always,

Your affectionate step-mother,

EMILY BUNBURY.

Tuesday.

Mildenhall,

August 3rd, 1847.

My Dear Katharine,

Fanny is not very well and knocked up by the heat of the weather and the excitement of the last few days, so I have undertaken to write to you in her stead.—The Bury election terminated on Saturday the 31st, and Edward came in triumphantly, having a majority of 65 over his opponent Mr. Twiss. The nomination was on Friday. We arrived at Barton on Thursday, and the next morning Fanny went into Bury with Lady Bunbury and little Harry, and my father and I went with Edward in a carriage-and-four, and were met at the entrance of Bury by a procession of voters on horseback and on foot, with banners and music, who escorted Edward through the principal streets, and so to the hustings. There were three candidates, but poor Lord Jermyn was very ill, and could not appear at the election, so his brother Lord 1847. Arthur spoke for him. Edward spoke extremely well, and Mr. Twiss made a clever lawyer-like speech. We were in Lady Bunbury's carriage, near the hustings.—Fanny and some other ladies in the carriage, and my father and I on the box, where we could see everything very well and hear tolerably. The polling began on Saturday morning, and at first Twiss's people made a push and gained a majority, but by twelve o'clock Edward passed him, and continued from that time to gain ground steadily and very rapidly after two o'clock. At four, when the election terminated, the numbers were,—for Lord Jermyn (I think) 398; for Edward 328; for Twiss 263. The whole thing went off very well, but there was much less noise and excitement than there was ten years ago, when I stood against Lord Jermyn. The display of red and white ribbons (our colours), was very gay, and Fanny looked extremely well in white, with knots of scarlet ribbons and a bouquet of scarlet geraniums and white roses. She was much knocked up, however, from sitting so long in an open carriage, exposed to the glare of the sun, which on Saturday especially was very powerful. Then on Sunday, which was another broiling day, she would go down to Bury to hear Mr. Eyre preach, and though she seems to have been much edified by the sermon, she suffered from the heat. Yesterday again was tremendously hot, and altogether by the time we arrived here, she seemed quite exhausted, but I hope the quiet of this place will soon set her to-rights.

1847.   Poor Cecilia was taken dangerously ill just before
the election, and you may imagine what distress
Lady Bunbury was in, and how much Fanny felt;
but I am happy Cecilia is now decidedly recovering.
She is staying at Bury, in the house of Dr. and
Mrs. Hake, who have shown her the greatest
possible kindness and attention.

I am glad to find that the elections, so far as they
have hitherto gone, are on the whole rather favourable to the present ministry, though they have been
very unlucky in having four of the ministers themselves unseated,—Macaulay, Charles Fox, Hawes,
and Hobhouse.   What can possess the good people
of Edinburgh to turn out Macaulay for an unknown
man like Mr. Cohen?

I hope you will not get into the middle of a war in
Switzerland, as it would be inconvenient and
uncomfortable, especially as I think you have no
belligerent turn; but matters in that country really
seem, if one can judge from the newspapers, to be
tending to throat-cutting.   I hope you will run away in
time.—We have decided on publishing my notes on the
Cape, before we go abroad, so the time of our starting
for Italy is deferred, and I hope we shall see you before
we set off.   If we do not get away before the winter,
I fancy we shall have to give up the Rhine, and go
through France.—We have had, and still have, a
splendid Summer, and our garden is in greater beauty
than it has been every previous year; I wish you
could see it.   The garden and arboretum at Barton,
too, are in full splendour,—I never remember such a
year for all kinds of vegetation.

You may probably have heard that we have a 1847.
little nephew staying with us,—my youngest brother's
son, a fine little fellow; he is a prodigious pet of
Fanny's, and very fond of her, and she has far more
influence over him than any one,—indeed I do not
think he much minds anybody else. She will miss
him very much when he goes to school, which he is
to do on the 18th; but he is just the boy for a
public school,—very social, active, hardy, and bold,
and though not fond of study, intelligent and
quick.

Skye is in high health and spirits, and as great a
favourite as ever.

Ever your very affectionate brother,

C. J. F. BUNBURY.

Mildenhall.
August 8th, 1847.

My Dear Mr. Horner,

I thank you much for your kind letter of
the 1st, and it was extremely gratifying to me to
find that you approve of my letter on the rule
established by the Guardians of this Union, and
that you concur in my opinions on that important
subject.

I am very glad to hear that you have got a house
that suits you, and in such an agreeable situation.
The description of it sounds very attractive. I shall
be very glad to visit you there, though I cannot
think, without pain, of the house in Bedford Place
passing into the hands of strangers—that house in

1847 which I have passed so many pleasant and improving
hours and which was the scene of my courtship and
marriage. I shall always look back to it with an
affectionate remembrance. I thoroughly feel the
pain which it must cost you all to part from what
has been so long a happy home; but I hope your
new home will be the scene of as much comfort and
enjoyment as the old one has been.

I wish you could have seen Mildenhall, this year,
for it has been in especial beauty (our lawn and
garden I mean) the flowers uncommonly abundant,
and the results of Fanny's labours and plans
developed most satisfactorily.

Edward's success was very gratifying, and I hope
that Parliamentary life will suit him. You will have
heard from Fanny that there was an expectation
just at last, that I should stand for the County, and
that there was even an idea of sending a deputation
to me, but too late. I was much pleased to find
that I am not yet forgotten, and that there is still a
feeling in my favour among the Liberal party in this
portion of Suffolk; I shall try to cultivate this,
especially after our return from Italy, and it may
perhaps afford me a favourable chance at the next
election. It is impossible to foresee what the state
of parties or the turn of the political tide may be by
that time; but if the chances do not seem too
desperate, and if my health and my worldly circumstances are as good as they are now, I shall certainly
try to beat up the quarters of the Tories. I was
very much pleased at Shafto Adair's success. It is
very odd and very unlucky that so many of the

Ministers should be turned out, and especially I wonder at Macaulay's defeat; it does no credit to the good town of Edinburgh. Lefevre's defeat too is awkward, as he cannot now be the head and Parliamentary representative of the Poor Law Commission. Have you heard who is likely to be appointed tó that post? Although the result of the elections may not give the present Government so great a majority as we could wish, still I flatter myself there is no chance for the Protectionists and High Tories, as the Whigs and Peelites together will easily be able to beat them, and I have no fear henceforth of Peel's taking a retrograde course. The present Ministry may be beaten, but the renewed ascendancy of Peel would be little less favourable to temperate and wholesome progress than that of Russell.

Katharine's letters from Switzerland are very interesting, and particularly so to me, as I know most of the scenes she describes, and I have a lively recollection of the wonderful scenery of the Via-Mala and the strange dark gorge of Pfeffers.

Fanny is well and sends her love. Pray give mine to Mrs. Horner and to our dear sisters, and believe me,

Your very affectionate son-in-law,

C. J. F. BUNBURY.

P.S.—If you see Mr. Brown, pray remind him that he kindly promised to give me some letters of introduction to Italian botanists.

                                         Mildenhall,
                                    August 24th, 1847.

My Dear Joanna,

I thank you very much for the extremely
pretty paper knife you were so good as to send me,
and which I prize highly, and will keep with great
care. The fame of Peter Vischer, and even his
appearance were not unknown to me, for my father
has repeatedly spoken of his very fine and admirable
works at Nuremberg, and has a small bronze figure
of him copied from the same statue which you
mention. Your little German tours must have been
very interesting and agreeable, and will have left you
a fresh store of agreeable recollections. I have
read with great interest Katharine's letters from
Switzerland, and they have set me longing to see
that country again ; it is now eighteen years since I
was there, but my recollection of most of the scenes
she describes is fresh and lively.

I was very glad to hear that you recovered your
favourite Polly ; the chances seemed much against
it, but I suppose you will now take greater precautions
against his vagrant propensities. A parrot was not
so likely to find his way home again, as Master Skye.
One of Fanny's pets has been indulging in a fine
frolic : very early this morning, soon after 4, Fanny was
awakened by a strange sound under our window, and
as soon as it was sufficiently light, she saw Punch,
the donkey, galloping round and round on the lawn.
I got up, and looked out, and saw him too. We
were in great alarm for our flowers, but on going
round the garden this morning, I was happy to see

that nothing seemed to have been touched ; it would 1847. appear that his only object was to amuse himself by racing round and round on the lawn (where the marks of his hoofs are plentifully apparent) and in and out among the shrubberies. I suppose he got tired of the monotony of the paddock. Fanny thinks he got out by squeezing himself through between the wires of the iron part of the fence.— Skye is flourishing and comical as usual. We had a short but very pleasant visit from Mrs. Young*, and fortunately had a very fine day to show her this place to the best advantage. Her little Emily is an uncommonly fine child.—We have heard nothing of Harry† since he went to school.

You must be in an awful bustle, moving the accumulations of so many years from your old home to your new. I like the name you have given your new home.—Rivermede sounds romantic and mediæval, and reminds one of Runnymede and Magna Charta, and it is descriptive too, and appropriate. I hope to see you all looking very snug and comfortable in it before we go abroad.

It happens most unluckily that Murray is out of town, so that I cannot get an answer about my book, probably before the middle of September.

My love to you, father and mother and sisters.

Ever your affectionate brother,
C. J. F. BUNBURY.

---

* His wife's cousin. Mrs Charles Baring Young.

† The son of Captain Hanmer Bunbury.

Mildenhall.

September 25th, 1847.

My dear Father,

1847.   I have finished your book, and cannot help writing to you to say how much delighted I have been with it. You will, no doubt, have abundance of praise from much better judges than me, but I must say for myself, and you know I am not a flatterer, that I have scarcely ever read a more delightful narrative, or one that has interested me more. I read it from beginning to end with unceasing pleasure, and only wished it had been three times as long. I am particularly struck with the beautiful clearness and simplicity of the whole narrative, and the liveliness of the sketches of character. I cannot help regretting that what appears to me so masterly, should not be given to the public ; yet at the same · time I can perfectly understand your dislike to expose yourself to the snarling of Quarterly Reviewers, and the ignorant impertinence of newspaper critics ; not that I, by the way, had any cause to complain of these latter gentry, but I speak according to what I have observed of their usual behaviour. Did you happen to see the article in the *Quarterly* on Lord Holland's Reminiscences ? It was the very quintessence of venom.

Your narrative is a curious illustration of the extraordinary want of system, of definite purpose, of information, and, in short, of statesmanship, in our successive governments, during that great struggle with Napoleon. It is truly wonderful how, with such Ministries and such allies, we ever managed to hold our ground against such a man. The Roman History has been assiduously crammed (at school and college) into the heads of our destined statesmen, but how few of them have known how to profit by it. I am very sorry to think that I shall not see you again before we go abroad; but I hope you will enjoy your winter in Wales.

Mr. Horner is going on well, but recovers his strength very slowly, and is a good deal depressed in spirits. Fanny was not at all well yesterday, but it was principally, I think, the effect of fatigue, and she was wise enough to go to bed early, and slept sound for nearly twelve hours. She is not strong, and she *will* overwork herself. The school goes on famously, ten or twelve new boys are coming next week, and Mr. Phillips thinks there will be seventy before the end of the year.

We are going to spend the latter part of next week at Hardwicke.

Pray give my love to Emily and Cecilia, and believe me,

Ever your very affectionate son,
C. J. F. Bunbury.

Mildenhall,
October 13th, 1847.

My Dear Mr. Horner,

You will be glad to hear that I am getting on prosperously with my book. Three of the little

1847. drawings which I have prepared to illustrate it, were
sent up to Murray last week, and put by him into
the wood-engravers hands, and he says I shall have
proofs of them by the end of this week.   Thus
the prospect of our departure for Italy is beginning
to assume something of a definite and tangible
reality.   We now talk of leaving home about the
11th or 13th of November, to proceed, in the
first instance to Rivermede, where we hope to
stay nearly a week, and where I hope we shall meet
you ; we must afterwards spend some days in
London, to make our final preparations, so that it
may be the 25th before we actually leave England
behind us.   Miss Nicholson left us this morning,
having stayed with us since Friday ; I am quite
charmed with her.   I had indeed often met her at
your house but by thus spending four or five days in
the same house with her, I have become better
acquainted with her than I could have in a month in
London.   She is indeed uncommonly agreeable.   It
was very kind of her to come such a distance to see us,
and I was exceedingly glad of it, on Fanny's account,
as it was both a great pleasure and a very good
thing for her, to keep up her intimacy with one of
her old and valued friends ; while it was a gratification to me to converse with a person so rational
and so highly cultivated.   Fanny went with her to
Ely this morning to show her the Cathedral, and to
see her off by the half-past one train ; but I could
not spare a day from my proofs and drawings to
accompany them.

I fear you must find the manufacturing districts in

a very uncomfortable condition, for the state of the money market seems to be deplorable. I confess I do not understand the Bank question, and the other great financial questions which are now so keenly debated in the newspapers and at public meetings, and on which (I observe) *The Times* and *The Daily News* take opposite sides. I think I remember that you approved of Peel's Bank Act. The question appears to me so difficult and complicated as scarcely to be mastered without a long and special study.

We have had lovely weather the last few days, and the tints of autumn on our trees are very beautiful. I am happy to say my dear Fanny has been very well lately, but I shall have to guard carefully against her overfatiguing herself in travelling, and especially just before starting.

I hope to see you in November in a comfortable state of health.

Believe me,
Ever your affectionate son-in-law,
C. J. F. BUNBURY.

S

CHAPTER III.

———

1847. [In December, 1847, Mr. Bunbury and his wife went abroad, taking with them her sister Susan. Most of the remarks in his journal on Sculpture, Architecture, Painting and Antiquities will be omitted].

=====

LETTERS.

———

Hotel Mirabeau,
Paris.
December 5th, 1847.

My Dear Father,

I wrote to you two or three days before we left London. We set off the day we had intended, Saturday the 27th, and crossed from Folkestone to Boulogne in very rough weather, wind and rain, so that, in the three hours that the passage lasted, we had time to be very considerably sick.

The next day was fine, though the wind still continued high ; in the morning Susan and I walked round the upper town of Boulogne, which I found to be more picturesque than I had been aware of ; at one o'clock we set off in a neat omnibus, and had a pleasant drive of about six miles to Neufchatel, to

which point the railway is at present open. We
found the railway carriages very comfortable, and
the arrangements good, and were conveyed without
any accident or annoyance to Amiens, where we
stopped for the night. Fanny and I went to the
cathedral the same evening, while vespers were going
on, and were very much struck with the grand effect
of the interior; we went again the next morning
before starting, and had a good look at the cathedral,
both inside and out,—and a glorious building it is,
in truth, one of the finest Gothic churches I ever
saw : the wonderful height of the nave, the grand
simplicity of its general effect, and the rich circular
windows with splendid coloured glass, fill one with
admiration. The front is criticised by connoisseurs,
but to my eye its effect is most noble and imposing.
—From Amiens we had a very comfortable and
pleasant journey by railway hither, traversing (at
least in the latter part of the way, from Pontoise) a
much prettier and pleasanter country than I at all
expected. We have been here now six days, in a
good hotel, very conveniently situated in the Rue de
la Paix, close to the Place Vendôme. The weather
is excessively wet and dirty, but very mild for the
time of year. Lyell had given me letters of
introduction to two of the most distinguished
scientific men of Paris, Elie de Beaumont, and
Adolphe Brongniart, both of whom have shown me
great civility, and given me much information, and
under their guidance I have gone through part of
of the exceedingly valuable and interesting collections
in the Jardin des Plantes, and in the Ecole des

1847. Mines.   We have received much kindness from Fanny's aunts, Mrs. Power and Mrs. Byrne, and their husbands, who have been long resident at Paris, and I have become acquainted with a very agreeable friend of Fanny's, Madame de Tourgueneff, the wife of a Russian exile who resides here.   She is a very pleasing person, and her husband is an uncommonly clever and well-informed man, remarkably conversant with the politics of all nations.

I hope to find a letter from you at Nice, and trust it may contain a favourable account, as well of your own health as of Emily's and Cecilia's.

We are all pretty well as yet, and have hitherto been lucky enough to escape the influenza (la grippe) which is excessively prevalent in Paris.

I will write to you either from Lyons or Marseilles.

Believe me ever,

Your very affectionate son,

C. J. F. BUNBURY.

P.S.—I understand that M. de Tourgueneff's book on Russia,—" La Russie et lés Russes," is very valuable, written with great ability, and full of information.   It might be worth your buying.

LETTERS.

My Dear Father,

I wrote to you from Paris last Sunday
week. We remained there some days longer than
we had at first intended, for finding that a voiturier
was not easily to be met with, and that it was not
a usual mode of travelling in France, we sat about
looking out for a carriage, and by the kind assistance
of Fanny's friends, we succeeded in obtaining a most
comfortable calèche in excellent condition with plenty
of room for ourselves and our luggage for 8oof. (£32).
The carriage was examined by our friends, who pronounced it to be perfectly strong, sound and
serviceable, and M. de Tourgueneff, who had long
known the coachmaker from whom we bought it,
vouched for his honesty and trustworthiness.
Hitherto we have found every reason to be satisfied
with our purchase. I must say, in passing, that
nothing could possibly exceed the kindness we
received from Fanny's relations, Major and Mrs.
Power, and Mr. and Mrs. Byrne, during the whole
of our stay at Paris ; they were never weary of
doing everything they could devise that could be
useful or gratifying to us.

Well, we set off from Paris on Friday, the 10th ;

1847. placed our carriage and our four selves in it, on the railway, and were safely conveyed to Bourges, where we arrived in eight and a quarter hours, including a halt of half-an-hour at Orleans. The next morning we saw the magnificent cathedral, with which we were very much pleased, especially Fanny, who has a passion for Gothic architecture, and has studied it carefully.

Having spent the morning in viewing the cathedral, we did not set off from Bourges till noon; we intended to sleep at Moulins that night, but the roads were so bad that by six o'clock we had got little more than half way, and despairing of reaching Moulins before midnight, we stopped at a little place called Sancoins (which I never heard of before) where we slept at an inn little better than a public house.

The next day we went on, crossed the Allier, got an excellent dinner at Moulins, and went on two stages farther, to Varennes where we spent the night at a wretched little inn. We were much diverted with the odd shaped hats worn by the country women in the district for some distance on each side of Moulins; I think you have travelled this road, and must have remarked those queer hats, turned up both before and behind, but behind particularly, curling quite over the top, and gay with black velvet and pink silk. After Moulins, the roads improved very much and everything seemed to become more civilized.

We had a pleasant day's journey to Roanne, having at first a distant view of the volcanic mountains of Auvergne, and afterwards enjoying very fine views of the mountains more to the east, 1847. between the valleys of the Allier and the Loire. We enjoyed still more our journey yesterday, when we crossed over the mountains between Roanne and Tarave. We were all delighted with the scenery of those hills, and I for one was most agreeably surprised, never having imagined that there was anything like such scenery between Paris and Lyons. It has a peculiar character of its own, not reminding me much either of the mountain country of Wales or Scotland or of the Alps, or of the Appennines, but it is very interesting even in winter, and must be charming in early summer.

We reached Lyons last night, and are lodged in an excellent hotel, Hotel du Nord ; to day we have been walking about the town, seeing the rivers and Quays, the Place des Terreaux, where Cinq Mars and his friend De Thou were beheaded; and the cathedral, which has most splendid painted glass. But the day has been so foggy that we have not been able to ascend Mont Fôurviers. This is the first day at all unfavourable, since we left Paris, hitherto we have been most fortunate in having continued fine weather without anything like cold ; a great contrast to this time last year. At Paris, on the contrary, we had only two fine days in ten that we remained there.

We are all well and have enjoyed our journey very much, in spite of three nights spent at bad inns, and now having got so far South, we have many objects of interest to look forward to, before we reach Nice.

1847.   I have not seen a newspaper since we were at
Paris, and the latest thing I read was the discussion
respecting the Bank and the Commercial Crisis.
We saw Rachel at the Théâtre Français, in
"Cleopatre," a new play, and admired her very
much, and we were much diverted by Déjazet at
the Varietés.   Fanny sends her love.

Believe me,

Ever your very affectionate son,

C. J. F. BUNBURY.

JOURNAL.

December 16th.

Lyons to Vienne.   The road hilly but not bad.
The country uninteresting until more than half-way
from St. Simphorien to Vienne, when ascending
a long hill partly wooded, we catch a distant view
of the snowy Alps, soaring above the haze; the
interesting hilly country, speckled with white houses
in a very Italian style.   A clump of pines most
picturesquely placed in the foreground, added to the
Italian character of the landscape.

Plenty of Juniper along the skirts of the wood
on this hill, and Helleborus fœtidus coming into
bloom.

Long descent between the walls of vineyards into
the narrow, fertile plain bordering the Rhone.

Vienne is a picturesque old town, of narrow, dirty
streets and high, irregular, heaped up houses, built

partly at the entrance of a narrow gorge between 1847. two extremely steep and bold hills—promontories of the sort of table land behind—on one of which is an old castle looking down on the town; a small rapid stream coming down the gorge, and throwing itself into the Rhone. A suspension bridge across the broad, strong majestic stream of the Rhone. A large, old square tower with mâchecouli sand corner turrets, on the opposite bank of the river, just above the bridge,—traditionally connected with the name of Pontius Pilate.

The cathedral is a very large, venerable, stately building, very much in the same general style with that of Lyons ; it stands on an elevated platform or terrace, and is approached by a broad and high flight of steps which add to the grandeur of its appearance.

The Roman Temple, called the Maison Carrée is sadly defaced and mutilated, its columns having in the middle ages been built into shabby walls of coarse masonry, and their symmetry quite concealed. It was growing too dark for us to see the museum.

All along the road we met numbers of heavily laden waggons, each with only two large wheels, coming with goods from Marseilles ; often a donkey harnessed in front of the horses. Horses' collars with a high peak like a horn, and two arched projecting bars curving out laterally, beset with small bells and covered in part by a sheepskin dyed blue. Oxen used in ploughing, and sometimes donkeys also, the donkey leading.

December 17th.

A lovely day; a sharp white frost in the morning.
—Vienne to Valence. An interesting journey down
the valley of the beautiful Rhone. Road good, and
not in general hilly, though in the first stage or
two there are a few steep hills. Much variety
produced on the hills on the left bank, sometimes
approaching close to the river, sometimes receding
so as to leave a wide intervening plain.

Corn, vines, and green crops; no hedges, but
abundance of mulberry and walnut trees. In
summer it must be beautiful.

A very little way out of Vienne on the south, in a
field close by the road, is a Roman obelisk, standing
on a square basement pierced with four archways,
like the arch of Janus quadrifons at Rome; the
whole of very solid construction, of large square
blocks of stone, very well put together, supposed to
be a sepulchral monument.

The road for some miles from Vienne, through
the narrow, flat, fertile plain bordering the river;
abundance of the yellow osier (salix vitellina), conspicuous at this season by its bright coloured twigs,
in the low grounds; and the usual characteristic of
French river scenery, long lines of tall, spiry poplars.
Afterwards, the river making a wide sweep, the road
quits it and ascends a long hill.

Hills on the left bank, high and bold, veiled
all day by a thin white haze, which, without taking
away the sharpness of their outlines, concealed all
the details of their surface, and added to their
apparent height.

Towards St. Vallient, the hills on the left bank 1847. again approach the river, and a little beyond it, a bold rocky headland juts out from them, leaving only space enough for the road between itself and the river, and crowned by the picturèsque ruin of an old castle called the Tour de Ponsas. Here, according to tradition, Pontius Pilate destroyed himself.*

Approaching Tain, the hills on the left jut out towards the river in another promontory, the hill which produces the famous Hermitage wine; its north and west sides excessively steep, rocky, bare and rugged, with only a thin sprinkling of box bushes and stunted oak brushwood; the southern more sloping and clothed with vines.

Drive by moonlight from Tain to Valence; we crossed the Isere by a suspension bridge; and we stopped at a hotel outside the walls of Valence.

December 18th.

Valence to Orange. Road not so good as yesterday, and weather much less favourable. The course of the Rhone continues constantly to skirt the bases of the western hills, which are much higher than those on the east side of the valley. The surface of the plain continues for a long way together, perfectly level, like the bed of a dried-up lake. The surface soil of a deep red brown, and apparently very fertile; but immediately beneath this, the whole soil seems to consist of loose rolled pebbles. There is extensive

---

* There are at least three different places, where according to popular tradition he destroyed himself.

1847. cultivation of the mulberry throughout this plain.

At Livrons, on the north bank of the Drôme, the precipitous limestone rocks have a singularly variegated appearance, the strata alternately ochrey yellow, and bluish grey, and this with extraordinary regularity.

We crossed the Drôme, a rapid stream, occupying but a small part of its broad, sandy, or shingly bed.

Between Loriol and Montelimar, the hills on the left approach the river, and show quite the character of Provençal scenery; they are low, but extremely rocky or stony, rough, arid and bare, only sprinkled with bushes of Box, and other low growing evergreens.

We drove through Montelimar, a town of narrow, dirty streets, enclosed by an old battlemented wall, and situated in the midst of the plain, half-a-mile from the river. We dined at the Hotel de la Poste, outside the town to the South.

We arrived at Orange very late, having been eleven and a half hours on the journey, besides the time we took to dine at Montelimar.

December 19th.

A very wet day. We went out through mud and water, in a deluge of rain, to see the Roman antiquities of Orange, which are indeed well worth making an effort to see.

The triumphal arch, situated a little way out of the town on the Lyons road, is similar in general form and construction to the arch of Constantine at

Rome, and (as well as I can compare from memory) 1847. quite as large.

The theatre, nearly at the other end of the town is a grand ruin. The wall at the back of the scena, truly gigantic, 121 feet high, 334 feet long, and 13 feet thick, built in precisely the same style, as the amphitheatre at Nismes of enormous blocks of stone, most accurately squared and fitted together, without cement. The material also I think the same, or a very similar shell limestone to that used at Nismes. Two different ranges of projections on the outside of the wall, to hold the masts which supported the awnings, as at Nismes, the coliseum, the theatre at Pompeii, &c. Internally the general plan of the theatre may be seen, though not much remains complete ; two or three of the lower ranges of seats are preserved, and a few of the steps ; and the corridors which ran round between the different orders of seats, are distinguishable. The ranges of seats, as usual, ascend up the side of a hill, against which the walls that back them are built. At the top of this hill was the castle of the Princes of Orange, which was destroyed by Louis XIV.

The interior of the great wall of the theatre seems to have been entirely lined with coloured marbles, of which innumerable fragments of very rich kinds, in great variety, have been found under the soil and rubbish. Numerous fragments also of white marble, beautifully sculptured ; some figures in relief, two Centaurs, a Victory, &c. in a very perfect style of art. One column still remains attached to the wall, and portions of a finely sculptured marble

1847 cornice. Rich marbles and elaborate sculpture seem to have been as profusely lavished here as on the great buildings of Rome.

Altogether this theatre gives one a magnificent idea of the wealth and splendour and importance of the colony of Arausio. I do not know whether the length of the scena and the diameter of the cavea may be equal to those of the great theatre at Syracuse,—probably not, but I never saw the remains of another ancient theatre that approached this in imposing grandeur of effect.

Went on from Orange to Avignon, the road traversing a wide, nearly level stony plain, in great part cultivated, but not apparently fertile. We are now in the region of the Olive, and of the great Italian reed (Arundo Donax), they were seen in considerable quantities, in several places in this day's journey. The Cypress also much cultivated, and often planted in long rows, having a singular and striking effect. Where uncultivated, the ground thinly covered with small evergreen bushes, and dwarf trees.

A very remarkable group of mountains visible at first to the south-east, singularly peaked and craggy; the wildness of their appearance heightened by the clouds which floated above them. Mont Ventoux hidden by clouds.

The road carries us almost round the city of Avignon, before entering it, giving us an excellent view of its old walls with their curious mâchecoulis.

All along the road in these four days' journeys, we met innumerable waggons, many of them very long and very heavily laden, but never with more than two wheels. Mules are used in drawing them, as much or more than horses, and they are very large and fine animals, often larger than the horses of the country. Very often a donkey is harnessed in front, followed by two or three horses or mules.

1847.

December 20th.

At Avignon. Employed all day in sight-seeing. The Roche des Dens, a bold mass of limestone rock, precipitous, on the side next the river, included in the line of the old walls, which is here interrupted, the precipice having been a sufficient defence in this quarter.—A fine view from the summit :— the cathedral and the enormous frowning gloomy towers of the Papal palace on the side of the rock ; beneath the dingy crowded roofs of the city, and its numerous spires ; the winding of the broad Rhone, with its new suspension bridge, and the old broken bridge of St. Bénézet, the town of Villeneuve on the opposite bank, with its fortress perched on a rugged grey rock ; the wide plain stretching away to the east, bounded by the hills above Vaucluse ; and to the north-east Mont Ventoux, covered with snow, dimly seen but looking grand and gigantic.

We went through the interior of the Pope's palace, and were shown the halls of the Inquisition, the chamber of torture, &c., but they are full of rubbish and workmen, as alterations are going on, so that it was difficult to realize to oneself the

1847. scenes acted there. A strange show-woman, a hideous little old woman, the very model for a witch so well described by Dickens in his account of this place. The great tower called the Tour de la Glacière, immediately adjoining that of the Inquisition, where, in 1793, upwards of 80 prisoners were massacred by the revolutionary mob, and their corpses thrown down to the bottom of the tower; long streaks of blood still visible on the walls. Traces of the crimes of a brutal and frantic mob, side by side with those of the more deliberate and long continued atrocities of a priestly institution. Great number of skeletons found in the Oubliette of the inquisition, when it was first opened. The hall where Pierre de Lude, the Pope's legate, in 1441, blew up the assembled nobles of Avignon, in revenge for the murder of his nephew. The great chapel with finely groined roof, now divided, and all its proportions spoilt by being converted into dormitories for the soldiers—for the papal palace is used as a barrack.

The cathedral adjoining the palace, chiefly remarkable for the almost classical (though not very pure) architecture of its porch, supposed by some to have belonged to an old Roman temple. A circular arch flanked by fluted Corinthian columns, and mounted by an entablature with mouldings in classical style, and by a pediment too high for correct taste.

A museum founded by M. Calvert, an excellent collection, most creditable to the town; a good collection in several departments, and in particular a most rich and interesting series of Roman remains, 1847. works of art and domestic utensils and implements, all found in this department. There are glass vessels in very great numbers, and of almost every variety of form ; metal mirrors ; needles and bodkins both of bronze and ivory ; dice for playing ; the top of a cavalry standard ; bronze swords and daggers ; hundreds of terra-cotta lamps ; a vast number of small bronze figures, chiefly of deities, some of very good workmanship. Except in the Naples museum, I have scarcely seen so rich a collection of the kind. There are also numerous Roman inscriptions and fragments of sculpture in marble. Two large bas-reliefs, one representing a sacrifice, the other a triumph, both found (as were a great many more of the antiquities) at Vaison near Orange. A *sella curulis*, represented in bas-relief in marble. An enormous amphora. Another portion of the museum is a very respectable picture gallery, comprising among many other things, good specimens of Joseph, Carle, and Horace Vernet, of Deveria, of Mignard, and of Sebastien Bourden. Several interesting drawings, among others, Vernet's original sketches for his pictures of French seaport towns. A collection of engravings, framed and glazed.

December 21st.

Avignon to Nismes. Ascending a long hill between La Bégude and Remoulins, I had an opportunity of botanizing. One of the most prevalent plants is the little scrubby, prickly,

T

1847. holly-leaved Kermes Oak (Quercus cocciferæ). Another Oak (Quercus pubescens) of the same group as our English Oak, but with leaves downy at the back, and retaining them very late, for they are not yet entirely withered, is abundant in particular places, forming brushwood, but not general. Observed the lavender, the Cistus albidus and Smilax aspera.

We crossed the Garden, a handsome river, by a suspension bridge, between Remoulins and La Toux and we went a little out of our way to see the Pont du Gard. This is the third time I have seen that glorious monument of Roman magnificence. We saw it to the best advantage, the day being brilliantly clear and fine; the sunshine striking full on the magnificent aqueduct gave a glow to the rich warm colouring of the stone, heightening its beauty, while it threw the shadows of the arches, with grand effect over the rocks on the other side. We spent three-quarters of an hour here with great enjoyment, rambling and climbing about the rocks, and viewing the aqueduct in various directions. The wild rocky hills about the Pont du Gard, are covered with the Ilex, mostly in a dwarf form as a bush or stunted tree, and varying as much in the shape of its leaves as in our gardens. The Kermes Oak is less plentiful here.

December 22nd.

At Nismes. A most beautiful day. We spent our time very pleasantly in seeing at our leisure, the admirable remains of Roman splendour, in which this town is so rich, especially the amphitheatre, and 1847 the Maison Carrée.—We ascended to the top of the Amphitheatre, and looked over the parapet, which remains in excellent preservation. Brackets projecting on the outside, on a level with the top of the parapet, with holes for the masts which supported the *velaria*. Many upper ranges of seats (those which were appropriated to the poorer classes and the slaves) in a very perfect state of preservation ; so also a few of the lower ranges, those towards the middle entirely destroyed, the podium very much ruined :—the grandeur of the style of building, especially conspicuous in the noble corridors ; vast size and admirable fittings of the stones.

While looking from the top range of seats at the distant landscape, we saw the white stream of vapour from the engine drawing a train along the railway. The present seemed brought into singular and striking contrast with the past by this sight.

The Museum of Zoology contains many specimens of remarkable animals from this part of France. Some interesting birds kept alive in a sort of aviary outside ; especially a pair of the white Egyptian Vultures (V. Perenopterus) which breed among the rocks near the Pont du Gard ; and many of the beautiful little Pintailed Grouse, or Ganga Cata (Pretrocles arenarius), a rare bird in Europe, inhabiting the stony plain of the Crau, to the south-east of Arles.

December 23rd.

Another lovely day. From Nismes to Arles. Approaching Arles, we crossed the lesser branch of the Rhone—itself a very fine and majestic stream,—by a suspension bridge, and we traversed the upper corner of the Delta. An avenue of uncommonly fine poplars (not Lombardy) by far the best trees we have seen since leaving Paris. We crossed the larger arm of the Rhone by a bridge of boats, and entered Arles.

The amphitheatre of Arles (which is described in my journal of 1842) is more perfect in the lower part than that of Nismes, much less so in the upper, the seats almost entirely gone, except a few of the lowest ranges.

The Church of St. Trophimus—curious sculptures of its porch ; fine cloisters.

The museum : exquisitely beautiful female bust, said to be of the Empress Livia, the nose unfortunately broken.

LETTERS.

Marseilles,
December 26th, 1847.

My dear father,

My last letter to you was from Lyons, written on the 15th. Our progress hither has been rather slower than we at that time expected, for we found so much to interest us at Avignon and Nismes that we gave an entire day to each of those places ;

and having no particular reason for hurrying, we 1847.
have taken time to see thoroughly the principal
sights of this interesting country. As far as Arles
we got on very smoothly and comfortably, without any
accident or difficulty, but the day we left it (the day
before yesterday) we met with a rather awkward adventure. We were going from Arles to Aix, and had
got to the middle of the third stage, when we were
induced to deviate from the high road in order to
visit the aqueduct of Roquefavour, which was
represented as not being very much out of the way.
The postillion himself, however, did not know the
road, which turned out to be unfinished, and indeed
hardly *made* at all,—scarcely more than traced out.
We lost our way several times ; the day closed
before we had got half-way to the place of our
destination, and we went on almost in the dark,
along the most dangerous-looking road I ever saw,
winding along the edge of deep ravines,—the road
itself very narrow and rough, and no parapet.
However, thank heaven, we got safe down into the
low grounds ; but there our driver again missed the
way, and in trying to recover it, fairly upset us into
a ditch. Most happily this occurred in the most
comfortable and convenient place that could have
been selected for such an event, the carriage was
safely lodged on its side against a soft bank of mud,
and we were none of us in the least hurt ; but we
had to stand in the cold while the driver gallopped
off to the inn of Roquefavour (which proved to
be ten miles off) to fetch assistance. There were
some houses near the very spot where we were

1847. deposited, but they appeared to be uninhabited.
At length a carter came by, and good-naturedly
roused for us the inhabitants of a neighbouring mill,
who showed us a great deal of kindness and sympathy, and did all in their power to assist us. By-
and-by the postillion returned, with men and horses;
the carriage was extricated from the ditch, and
proved to be perfectly uninjured, except that the
lamp glasses were cracked. It was too late, however,
to think of going on to Aix, so we proceeded merely
to Roquefavour, and slept at the little inn, or rather
public house, there, where we found the people
exceedingly civil and obliging. So ended our adventure. The first time in all my travels that I ever
was overturned. I was very much afraid that
Fanny would have been quite ill with the fright,
fatigue, cold, and hunger, for she had eaten nothing
from eight in the morning till ten at night,— but
happily she seems to-day to be quite well, and we
are none of us a bit the worse for the adventure.
The aqueduct of Roquefavour which we had gone
through so much to see, is indeed magnificent, the
grandest modern work, I think, that I have ever
seen. It is an imitation of the Pont du Gard, but
surpasses it considerably in size, being 272 feet high,
and 1230 feet long at the top ; it has not however so
massive an appearance, the arches of the middle tier
being much higher in proportion, both to their own
breadth and to the thickness of the piers, which makes
these piers, though really very massive, look comparatively slender, perhaps almost to a fault. The masonry
appears admirable. This great work was seven years in

building, and was finished about six months ago ; the 1847 engineer and architect is M. Montrichet, quite a young man. It does honour to France. The object is, to convey water from the Durance to Marseille. Soon after passing Salon, on the road from Arles to Aix, before we turned out of the high road, we saw the first pine woods, and from thence they became conspicuous features of the scenery, and very beautiful they are. They consist entirely, so far as I have yet seen, of the Pinus maritima of Lambert, which the French botanists refer to Pinus Halepensis, but it seems to me as it did to you, to be not exactly the same as the cultivated Pinus Halepensis, perhaps however, it is hardly a distinct species. We have enjoyed very much our journey from Lyons, and especially the sight of the magnificent Roman remains at Orange, Nismes, and Arles. Those of Nismes I had seen twice before, but I was as much pleasd with them as ever, and had additional pleasure in seeing how much they interested and delighted Fanny. Those of Orange were new to me, and are very striking, especially the grand remains of the ancient theatre. We were very much pleased also with the valley of the Rhone, from Lyon down to Montelimar, —but as you know all that country well, I need not attempt to describe. I do not know whether you ever saw the Museum at Avignon : it is an interesting collection, very creditable to its founder, and to the town, and in particular the collection of Roman implements and works of art, small bronzes, &c., is extremely rich and curious. Except at Naples, I hardly remember to have seen a more interesting

1847. collection of the sort ; and these objects were all found in the department in which Avignon is situated. —In the Maison Carrée at Nismes, the interior of which has been turned into a museum, we saw the best picture of the modern French school that I have ever seen,—Paul Delaroche's "Cromwell opening the coffin of Charles the First." We arrived here yesterday, Christmas day, but as the weather was excessively stormy, raining, and blowing violently, we saw nothing of the beauty of the scenery. I long for the letters from home which I hope are awaiting me at Nice, and I trust to have good accounts of you and Emily and Cecilia. I am very anxious also to hear what you think of my book.

Fanny sends her love. Believe me

Ever your very affectionate son,

C. J. F. BUNBURY.

Marseilles,
December 26th.

My Dear Mary,

We have come so far on our journey to Italy, having travelled leisurely, especially since we left Lyons, and taken time to see well the very fine and striking Roman remains and other curiosities of this interesting country.

But I know that you will have received from Susan most copious and accurate accounts of all that we have seen and done, for I have constantly observed and admired her indefatigable perseverance in writing letters every day of our journey, and

under all circumstances; so I will not go over the 1847.
same ground, but try to supply some particulars
which may not be quite so much in her province.
To her also I must refer you for an account of our
adventure of the night before last, as I have just
been writing about it to *my side of the house*, and do
not want to go over it again. It was the first time
in all my travels that I ever overturned; indeed it
was the most like a serious adventure such as one
reads of in old books, of anything I ever met with.
Susan and Elizabeth behaved really like heroines;
Fanny went through it bravely enough at the time,
but as soon as we were safe in the little inn of
Roquefavour, she became quite ill from alarm and
fatigue and exhaustion (she had been fasting almost
entirely for fourteen hours), and I was very anxious
about her, as she continued in a state of painful excitement and suffering all night; the next morning, however, she was much better, and after a night's rest here
she seemed quite restored, and I trust will not be at
all the worse for our adventure. It was certainly a
great mercy that we were not thrown over a precipice.

The aqueduct of Roquefavour, which we went
through so much to see, is really a magnificent and
surprising work, and does honour to France; a
modern aqueduct, ninety feet higher than the famous
Pont du Gard, and above 300 feet longer, and constructed altogether in the most admirable style. It
was seven years in building, and has been finished
about six months.

I am delighted to hear of dear Katharine's

1847. approaching happiness, and hardly know whether the bride or bridegroom is most to be congratulated. I was particularly pleased with Captain Lyell, when we were staying in your house, just before we came abroad. I liked his countenance, his manners and his conversation, and I should think, as far as one can form a conjecture on such momentous points from so short an acquaintance, that he is a man who would make a very good husband, and I am sure Katharine deserves one. Her letter to Fanny which we received here, is delightful. I hope they will. be as happy as Fanny and I are. I am sorry that we cannot possibly be present at the wedding. It is a satisfaction that they will not be obliged to remain very long in India, and I trust there will be no more wars. I have had no time for geologizing on our journey beyond observing in a cursory way the general nature of the rocks ; and the season is not the most favourable for botany. But there is something very striking and interesting to observe, in the complete change of character of the vegetation, when one enters these southern regions. This is hardly apparent till we reach the region of the Olive, which begins about Avignon, or a very little to the north of it ; for though, all down the valley of the Rhone, even from Lyons, one remarks the general cultivation of the Mulberry (for feeding the silk-worm) as an exotic feature in the landscape, it is not accompanied by any striking peculiarities in the native vegetation. But as soon as we have passed Avignon, the peculiar silvery-grey of the olive begins to predominate in all. the cultivated parts of the landscape, accompanied

in the lower grounds by the equally characteristic 1847
form of the great tall Italian reed, Arundo Donax ;
and wherever the hills remain in a state of nature,
we see the peculiar features of the Provençal
vegetation,—the little scrubby, prickly-leaved, holly
like Kermes Oak, the Quercus coccifera, which forms
the hardest and most rigid of bushes, scarcely a
yard high, the Lavender, a very aromatic grey-
leaved kind of Thyme, different kinds of Cistus, a
Juniper different from the English one,—and in
sheltered situations, the climbing prickly Smilax and
various other plants unknown in England except in
gardens. The hills about the beautiful Pont du
Gard, are covered with the evergreen Oak, intermixed
with the different shrubs which I have just mentioned. There is something in the general aspect of
this vegetation, which reminds me very much of the
Cape of Good Hope,—taking for the comparison an
equally dead season of the year ; the plants themselves, indeed, are all different, but there is a
striking similarity in the hard, dry rigid, evergreen
character of the bushes, the grey woolliness of many
of them, and the way they are scattered over the
arid rocky ground.

In our journey of the day before yesterday, a little
before we deviated from the high road, we saw the
first pine woods, which soon became predominating
features of the country, the hills about the aqueduct
of Roquefavour are covered with them ; they are
composed of the Pinus maritima or Halepensis, not
one of the largest kinds of pine, but one of the most
elegant and graceful, much less rigid than the

1847. generality of pines, with slender delicate bright green leaves. Under these trees the ground is covered with the little Kermes Oak, with Rosemary, Lavender, Thyme, Cistus, two kinds of Juniper, and the Provençe Furze. These pine woods (in one of which we lost our way for some time at the beginning of our adventure) are charming.

All the hills of this olive and pine country, consist uniformly, so for as I have seen, of one kind of lime-stone, scarcely varying at all except in colour ; the soil of the valleys and plains is generally full of rolled stones ; and to the eastward of Arles we traversed for several miles the singular plain called the Crau, a perfectly dead flat, covered uniformly with loose rolled stones like the shingle of the sea-shore. It is altogether a singular and interesting country, and would be well worth a visit from you and Charles Lyell,—though I suppose its geology must have been pretty well worked out by the French savans.

I must leave off now, to leave a corner for Fanny, so fare you well.

Ever your very affectionate brother,
C. J. F. B.

JOURNAL.

Marseilles.   December 26th and 27th.

Marseilles to Brignolles.   December 28th.

Brignolles to Fréjus by Le Luc Vidauban and Muy.

The ruins of a Roman amphitheatre seen on the left hand, not far from the road, a little way before entering Fréjus.

Magpies amazingly numerous in this country.

————————

December 30th.

Fréjus to Cannes. Immediately after leaving Fréjus, saw considerable remains of a Roman aqueduct on the left, close to the road, mostly single piers and broken arches, but here and there two or three arches together remaining perfect. It is built of small stones, not very regular, with much cement between them,—very much in the style of Caracallas baths,—the arches of thin Roman bricks.

The morning was unfortunately very wet, so that the beautiful Estrelles mountains were wrapped in clouds when we crossed them, and much of their grandeur was lost to us.

Arrived at Cannes, charmingly situated in the curve of a beautiful little bay, between two chains of mountains, which run down to the sea, and form the two horns of the bay.

Numerous pretty villas near Cannes, among them, Lord Brougham's.

We took a delightful walk in the evening along the sea-shore ; the sea gently rippling, and breaking with a pleasant sound, and a delicate line of white foam on the sands. The sunset glorious ;

1847. the beautiful chain of peaked and cragged hills,
which form the western boundary of the bay,
at first almost lost in the splendour of the sun;
then, as it sank beneath them, appearing of an
intensely deep blue, then changing to a dusky purple,
and finally almost black; while the clear sky above
them (over where the sun had set) was of the most
glowing yellow, softly melting off into the deep
purple blue shore; and in the opposite quarter, the
clouds exhibited an indescribable variety of beautiful
colours.

There is a very decided tide here, its traces were
clearly perceptible on the wet sand, and though I
cannot say precisely what the rise may be, I think
it cannot be much less than three feet perpendicular.

The rocks which appear on the seashore at Cannes
are of gneiss, composed principally of red felspar,
and black mica, the latter in great quantity; the
laminæ in many places curiously twisted and
curled.

———

December 31st.

Cannes to Nice.   Island of St. Marguerite,
opposite to Cannes, where the Man in the Iron
Mask was imprisoned.

From Cannes to the Var, the drive is through
one continued garden of olives, intermingled occasionally with orange and fig trees.   The bright
blue sea on the right,—on the left cultivated hills
crowned with pines; in front, a long line of finely-
indented mountainous coast, and the beautiful snow-

clad range of the maritime Alps rising over all. The road occasionally skirts the sea ; but rises for the most part over a succession of low hills, commanding from every height lovely views of the sea and coast. The town of Nice conspicuous from a distance, white and glittering.

The Olive trees here are very large and fine ; their trunks excessively thick towards their base, rugged and cavernous and twisted in the most fanciful way.

Picturesque groups of country people gathering the Olives.

The peasant women have remarkable hats—almost Chinese.

We had a long detention for want of horses at Antibes, a regularly fortified town, in a very pretty situation. A strong fort on a small peninsula near it.

The Var is a very rapid and impetuous torrent-river with a very broad shingly bed, of which the greater part is dry. A long wooden bridge over it, with the French passport office and custom-house at one end of it, and the Sardinian at the other.

From hence to Nice the road runs through a low and marshy but cultivated plain, skirting the sea shore ; the Italian reed, cultivated in great quantities along the road-side and the borders of the fields ; here and there the American Aloe.

We passed through a long suburb, and entered Nice.

So ends the year 1847. God grant that the next year may be well spent.

LETTERS.

My dear father,

1848.   We arrived here on the last day of the year, and it was a great pleasure to me to receive your letter of the 18th. I am very happy to hear that you have got over the influenza, which seems to have been so formidable in London, and I trust you will have no relapse, but will enjoy good health throughout the year which is now begun. It is a great pleasure and satisfaction to me to know that you like my book. I hear that 320 copies of it were disposed of "to the trade" at Murray's annual sale, and that this is considered a good and satisfactory beginning. I am sorry that Murray has put me into sky blue, which besides being the Tory colour, is I think too gaudy for any book except an Annual. I have as yet heard no other opinions of the book, but there can be none more important or more valuable to me than your approbation. Sir George Napier found us out and came to see us, full of his usual frankness and cordiality, almost as soon as we were lodged in our hotel; he is in high force, and so is Lady Napier. I never saw either of them looking better. Yesterday we saw Sir Charles Napier and all his party, who are settled here, I believe for the

winter; he was in excellent spirits and looking very 1848.
grand and singular, with his great nose and eagle
eyes, and immense beard and moustachios. William
Napier is very flourishing, and very little changed
(to my eyes) since I saw him seven years ago,
though he is a husband and a father. His wife has
an interesting countenance and pleasing manners.
I was very glad to become acquainted with poor
John Napier's widow, and was much pleased
with her. Yesterday Sir George Napier induced
me to go with him in the morning, to Government
House, where there was a "reception," being New
Year's day, and presented me to the Governor; and
in the evening, after dining with him, we all three
went with him and the whole of the Napier party to
a very crowded conversazione at the Governor's—
which I thought tiresome enough. On both
occasions Sir Charles Napier was the great
object of attention and observation — indeed he
is prodigiously *fêté* here. We were very sorry to
find that we had missed the grand ball given in
honour of him by the English at Nice, for which we
were just a few days too late; we heard of it for the
first time (after it was over) at Cannes, and Fanny
was quite in despair. I am very glad to find
however, that Sir Charles is properly appreciated by
his countrymen here, whatever may be the case in
England.

Nice is very full at present, and besides the usual
abundance of English, there are great numbers of
Russians. We were not able to obtain rooms in
the best hotel, which is kept by your old acquaint-

U

1848. ance Buonaccorsi, but we are very tolerably well off
in the Hôtel de Londres, in an excellent situation,
and very near to the Napiers. I hope you will
have received my letter of the 26th December from
Marseilles in which I gave you an account of our
night of adventures, and our over-turn when in search
of the aqueduct of Roquefavour. We were none of us
much pleased with Marseilles; indeed I reckon it an
error in my arrangement of our journey, that we
went there at all, but being there, we were obliged
to wait till the 28th, for the *visa* of the passport
and other necessaries. I do not wish ever to see
the place again. However on the 28th we got away,
and came hither in four days, sleeping at Brignolles,
Fréjus and Cannes. This journey was through a
very pretty country, but we were unlucky in having
a very wet morning to set out from Fréjus, so that
the beautiful Estrelle mountains were wrapped in
clouds, which concealed from us much of their
scenery; we could however see and admire the
beautiful evergreen woods which clothe them, and
Fanny was delighted, as I expected she would be,
with the Myrtle and Mastic and Arbutus, and other
fine shrubs, which she had never before seen wild.
We all admired the Cork trees very much. It is
singular how different they are from the Ilex in the
general form and character of the tree, while in
herbarium specimens they are so like that it is difficult
for any botanist to distinguish them.

Since we arrived here we have learned that
a carriage was stopped and robbed on the
Estrelles not very long ago. I am glad we

did not hear of it before, as Fanny might have 1848. been alarmed.

We were all charmed with Cannes, its hotel and its scenery. The rain had fortunately ceased before we arrived there, and we enjoyed a ramble on the sea shore in a lovely evening. I was surprised to find that there is a very decided ebb and flow of the tide there. The next day was delicious, and Fanny and Susan for the first time saw the *blueness* of the Mediteranean, which hitherto had shown itself to them in colours not different from those of our channel. The weather has ever since been as fine as possible, cloudless skies, a powerful sun, swallows and butterflies flying about (yes, literally), and a temperature, such that I *voluntarily* open the window, are not much like the beginning of January. We hear however that there has been a great deal of rain lately, and indeed the state of the roads shows it.

I beg you to thank Emily for me for her letter, written on Christmas day which I have received this day, January the 3rd, but I am sorry to find from it that both of you are so slowly recovering from the influenza, and that you have such dreary uncomfortable weather. I have caught a cold which is rather troublesome, but which does not in the least interfere with my appetite or my power of walking, and all our friends here say that both Fanny and I are looking remarkably well; indeed I think we are all much the better for our journey. We propose to remain here till about the 14th, then to go to Genoa.

1848.   *(January 5th).*   We have been out so continually,
that I have not yet had time to finish this letter.
We see a great deal of the Napiers and find their
society exceedingly pleasant ; and then the weather
is so delicious that we are tempted to be out all day.
We are all charmed with Nice.   Yesterday, Fanny
and Susan and I made a delightful expedition (they
on donkeys and I walking) to Villa Franca ; and to-
day we joined in a grand party with Sir Charles and
the William Napiers and Mrs. John Napier, and
Mr. Moysey, to Saint Andrea, and went on beyond
the convent, by a scrambling difficult path to a
pretty little dripping cave hung with the most
beautiful drapery of Maidenhair Fern and Moss that
I ever saw.   Did you go to this grotto when you
were at Nice ?

Pray give my love to Emily, and with most cordial
good wishes for the new year to both of you, from
Fanny as well as myself, believe me,

Ever your very affectionate son,
C. J. F. BUNBURY.

Nice,
January 9th, 1848.

My dear Edward,

A few days after our arrival here, I received
your interesting letter written on Christmas Day,
for which I thank you very much.   I hope with all
my heart that you may enjoy health and happiness

through the year which is now begun, and through 1848. many more after it. I am very glad that you find your parliamentary duties so interesting, but I pity you for having to go through the insufferable bore and worry (at least it would be so to me), of changing your abode, after you have so long been domiciled in Lincoln's Inn Fields, and have accumulated such a store of "goods and chattels" around you. We arrived here on the last day of the old year in lovely weather,—the sky cloudless, the sea brilliantly blue, the swallows flying about, everything looking and feeling like spring, and so it continued for the first five days of our stay ; since then it has changed, and we have had dull, gloomy weather, rain here and snow on the hills not far off —but as yet no severe cold. We found the whole party of Napiers, a pretty numerous one, established here ; Sir George and his wife in one house, Sir Charles and *his* wife, and poor Mrs. John Napier, in another ; the William Napiers in a third ; all very near to the hotel in which we are lodged, and as you may suppose, we have seen them almost every day. Sir George is in high health and spirits, and as full of hospitality and warmth of heart as ever, and speaks as good French as ever. Sir Charles, with his great nose and his great beard, has very much the air, when in plain clothes, of a venerable old Jew. D'Israeli would be delighted with him, he looks so like one of "The great race ;" but in uniform he is magnificent. He is in excellent spirits, extremely entertaining and agreeable, and full of good stories about India. He is prodigiously

1847. admired and *fêté* here, but the country people look at him with astonishment, and some of them cross themselves as he goes by. Besides the Napiers, there are here of our acquaintance, Mr. and Mrs. Moysey, who are connections of Mrs. Richard Napier, and very pleasant people ; Mrs. Moysey is very pretty.

We made out our journey through France, very satisfactorily, and I must say that I came away with a more agreeable impression of that country, and even of the people, than I ever had before. It was certainly fortunate that we took it in our way to Italy, and not in returning, for we enjoyed much that would seem tame and insignificant after the scenery and towns of Italy, and everything kept improving on us as we came southward, whereas all to the north of Lyons, would have been very wearisome if we had been travelling the other way. Our carriage has been a great comfort. We were agreeably surprised to find the road from Lyons along the left bank of the Rhone very good, at least as far as Orange, whereas we had been repeatedly told that it was one of the worst in France,—and so Murray says. In fact from Lyons southwards we met with no seriously bad roads, till, between Arles and Aix, we deviated from the beaten track to visit the newly-finished aqueduct at Roquefavour ; and then we met with an adventure which might have ended tragically. The postillion, though he had induced us to make this détour, did not himself know the road, which turned out to be unfinished, and in parts very dangerous, winding along the edges of precipices without any parapet. In all my travels I do not remember to 1848 have seen any road more alarming than some parts of this, along which we crept in very dubious twilight. At last we were fairly upset, but happily in the most safest and most convenient place, where such an event could have happened, the carriage being safely deposited on its side against a nice soft bank of mud, and neither it nor we at all hurt. However, there we had to stay in the cold and nearly in the dark, till the postillion fetched assistance from Roquefavour, which was ten miles off, and by the time we reached that place, it was too late to go any further, and so we spent the night there, in a very poor little inn indeed, but the people were very civil and obliging. The next day we went on to Marseilles. The aqueduct of Roquefavour is certainly a surprising work for the present day, and well worth seeing, but if ever you go to see it, I would recommend you to go from Marseilles which may be done without any difficulty or risk.

The aqueduct is similar in its general plan and construction to the Pont du Gard, but is both higher and longer,—not quite so well proportioned, however, for the arches of the middle tier appear too high, relatively to the other dimensions. It was finished about six months ago, after being seven years in building.—We took our time in travelling through that interesting country in the south of France, and saw Nismes and Arles thoroughly and to our full satisfaction. Fanny as much delighted with the Roman antiquities as I had hoped she would, and though I had repeatedly seen them

1848. before, I enjoyed them as as much as ever. Fanny's antiquarian zeal and ardour for exploring reminded me of you, especially when she kept me shivering in the cloisters of St Trophimus at Arles, exactly as you did in '42. Seriously however, that part of France is remarkably full of objects of curiosity and interest; and in particular I was delighted with the very rich collection of Roman antiquities in the museum at Avignon, which I had not seen before.

It appears very singular that while Provence generally is so rich in remains of classical antiquity, Marseilles, a place so early and so long famous and of such importance in ancient time, should have retained no such relics.

I am much pleased with the situation and scenery of Nice, and with the surrounding country, but I am not sure that I should like it for a long stay—it seems too much a place of idleness and frivolous dissipation. We made two very pleasant excursions while the fine weather lasted, one to Villa Franca, and the other to the beautiful grotto of St. Andrea. We originally intended to stay here no more than a fortnight, but I do not think we shall get away while the weather remains unsettled, as I have no fancy to be stopped by swollen torrents on the Corniche road. We think of staying a month at Genoa, if we can get good lodgings for that time, and then going on by Pisa and Florence without stopping for more than a week at either, to Rome; deferring our fuller acquaintance with Florence, till our return northwards.

I hope you like Captain Lyell; I did, extremely 1848. when we were staying with him at Charles Lyell's, before we left England. Katharine seems to be very happy, and everything going on as well as possible. We hear that the wedding is fixed for the 25th.

Fanny sends her love, and hopes you received a letter she wrote you from Avignon.

Ever your very affectionate brother,

C. J. F. BUNBURY.

JOURNAL.

January 10th.

The little plain of Nice is so richly cultivated as to look like a continued garden; with the exception of the plain of Rio de Janeiro,* I have never seen anything richer or more smiling. The lower hills are covered in every part with plantations of olives, which extend also up the principal chains of hills on each side of the plain or valley of the Paglione to a very considerable height, and in some places almost to their tops. The sides of the hills, even where most steep and rocky, and barely accessible, are cut into terraces, faced with rough stone walls and planted with olive trees; and in many places these are seen flourishing among the

---

* I ought to have excepted also the "Couca d'oro," the plain of Palermo, which has quite a semi-tropical character.

1848. rocks, and even intermixed with the pine trees,
which are the native growth of the hills. In the
plain, too, the olive predominates over the other
cultivated plants, though not to so great a degree as
on the hills.

The gardens in and around the town, and throughout the plain, up nearly to where the Paglione issues
from the hills, are full of Orange trees, which grow
luxuriantly, and show the genial nature of the
climate by their vigorous, deep green foliage, and the
profusion of fruit which they bear. They give a
rich and beautiful effect to the plain when one looks
down on it from any of the nearer heights.

The sweet bay grows here to a large size, really a
tree, twenty feet high and upwards.

The American Aloe is seen in many places
flourishing vigorously in the crevices of the stone
walls which bound the olive grounds; and on the
Castle Hill at Nice there are some plants of it which
have lately flowered. The prickly Pear also thrives,
but seems to be a less frequent object of cultivation,
—I have seen it only in two or three spots. There
are Date Palms in some of the gardens in the town
and its outskirts, but of no considerable size. I
have noticed also in several of the gardens, and on
the Castle Hill, a large arborescent Cassia, a
thoroughly tropical form, growing vigorously and
flowering freely in the open air. The vine is of
course extensively cultivated, but is not conspicuous
at this season, and seems in reality to be quite
subordinate to the Olive. The Arundo Donax is
one of the most conspicuous plants of the low

grounds, where it forms extensive beds, and grows to 1848. a surprising height, and contributes as much as almost anything to the exotic and southern aspect of the scenery, but it is cultivated also here and there, at considerable elevations on the hill sides, in dry, stony ground, where, however, it is by no means of so large a growth.

The most frequent and conspicuous of the native trees, growing plentifully on the arid, rocky tops of many of the higher hills, and clothing the sides of some of them which are too steep for cultivation, is the Maritime Pine (Pinus Halepensis),—the same which we saw in such abundance at Roquefavour, and on the road from Marseilles hither. When growing among the arid limestone rocks, with little soil, it often becomes exceedingly irregular and fantastic in its growth, with a singularly crooked and twisted stem. In some instances, on the exposed sunburnt tops of the hills, I have seen it assuming a straggling bushy form, almost like the Pinus Pumilio. The Pinaster grows in company with it, but is much less plentiful.

Among these pines, and in the stony uncultivated spaces between the olive grounds, on the sides of the hills, grow many of the characteristic forms of the Provençal vegetation, the Cistus albidus, Juniperus oxycedrus, Rosemary, Thyme, and a great variety of dwarf aromatic undershrubs of the Labiate tribe. But I have not seen here either the common Lavender or the Kermes Oak, which are so characteristic of Provence,—nor is the Ilex abundant. The Mastic grows plentifully in these situations,

1848. the prickly twining Smilax forms graceful wreaths among the rocks and bushes, and the Myrtle occurs here and there, but I have not yet seen it in any considerable abundance. The Caruba tree— Ceratonia Siliqua—is frequent on the rocky, sunny hills, and grows in some places to a large size ; there are some fine trees of it near the road in the descent to Villa Franca.

Three species of Euphorbia are now in flower here : one, Euphorbia segetalis, the smallest of the three, with yellow flowers and narrow, glaucous leaves, resembling those of some kinds of Toadflax, is common everywhere in the olive-grounds. The Euphorbia Characias, a fine, tall showy plant of its kind, is abundant in many places hereabouts, among the limestone rocks and on the ruins of the Roman amphitheatre. The third species, Euphorbia den- droides, grows in profusion on the precipitous rocks of the Castle Hill, and is pretty plentiful also on the arid, rocky hills above Villa Franca, forming large round compact bushes, a yard high, of a fine glaucous green colour, or a bright yellow when in flower. Its mode of growth is very peculiar : the stem, which is very thick in proportion and quite woody, like a miniature tree, divides into several branches arranged in an umbellate form, each of which is again repeatedly divided and subdivided in a similar manner. The Arum arisarum, with its shining green leaves and curious brown-striped hooded spathes, is exceedingly plentiful under walls by the sides of the lanes and roads, and also among fragments of limestone rocks. It is often intermixed

with the Arum Italicum, which is not yet in flower, 1848.
but may be recognised by its large leaves veined
with white.

The only grasses I have observed in blossom are
two species of Andropogon, a genus very characteristic of warm climates.

The most common Fern here is the Ceterarch offici-
narum, which abounds on all the rough stone walls
and in the crevices of the calcareous rocks. The
beautiful Adiantum does not seem to be at all
general in this neighbourhood.

January 21st.

Dr. Adolphe Perez shewed me his interesting
collection of fossils from the neighbourhood of Nice.
He tells me that the limestone which prevails in
this neighbourhood is of the Jurassic period, and
that he can distinguish by the fossils three divisions
of it, corresponding to the Great Oolite, the Oxford
Clay, and the Lias; but there are no lithological
differences. Fossils are difficult to meet with except in particular localities, but he showed me many
different species of Ammonites, Crioceras, and
other Cephalopoda, which he had procured from
these rocks; several of them identical with English
species, particularly characteristic of the above-
mentioned formations. Above these rocks he re-
cognises a limestone corresponding with the Neocomian (lower green sand) formation, and containing
like the Neocomian beds of our own country, dark
green grains; and still higher beds belonging to the
Cretaceous period.

1848   The clay which underlies the conglomerate of the
Vallone Oscuro is Pleiocene, and contains innumerable minute Foraminifera.

Dr. Perez holds that the Nummulite limestone
which occurs on the Riviera, about Ventimiglia,
San Remo, and many other places, but not
nearer to Nice than Mentone, is *Eocene*, not
Cretaceous. He showed me abundance of the
Nummulites, and also numerous species of shells
which he had procured from the beds associated with this Nummulite limestone, and which
he identified with Eocene species (among others
Crassatella tumida), but he said he had found
in the same beds *one* shell of the Cretaceous epoch.

This formation of Nummulite limestone contains
also the Fucoides intricatus and Targionii, which
occur in the south-west of France in strata supposed
to be Cretaceous.

January 22nd.

Left Nice at nine o'clock,—the morning lovely.
Very long ascent, the road rising gradually up the
side of the mountains, on the left of the Paglione,
in a direction nearly parallel for a considerable
distance to the Turin road. Delightful views of the
beautiful and rich valley of Nice, with the innumerable gay white houses and convents dotted over it,
the pretty town with its castle rock, and the
succession of bays and headlands to beyond Cannes.
At length, winding round the tops of the hills, we

look down from a great height on the sea, the 1848.
peninsula of Saint Ospizio, Villa Franca with its
beautiful little bay, and the promontory of Mont
Albano. The road runs along the brow of the
mountains, which rise with extreme steepness from
the sea. Much snow along the side of the road,
and on the hill sides, both above and below it.

Esa a curious looking little old town, in a most
singular and picturesque situation, clustered on
the sides and top of an abrupt pyramidal rock,
apparently almost inaccessible—to right of the road
and below it.

Immense cliffs of limestone rock, stratified with
singular distinctness, towering high above us, on the
left, and forming the crests of the mountains,—below
on the right, the descent almost precipitously steep,
to a vast depth. The mountain sides in general
quite bare, bristled with projecting crags; here and
there, scattered pines springing from among the
rocks; and in many places, portions of ground
terraced and cultivated, with extraordinary industry
even amidst the almost perpendicular cliffs.

Euphorbia dendroides very abundant among the
limestone rocks on this coast, forming a very handsome bush. Euphorbia Characias also in several
places. Turbia, a village standing very high on the
mountains; the ground all about it covered with
snow. A great tower-like mass of solid masonry
rising very conspicuously high over it,—the remnant
of the Tropea, the triumphal monument erected by
Augustus to commemorate his conquest of the
Alpine Tribes.

1848.   Monaco, visible far below us, looking like a neat
little model, or toy town, on its little promontory
curving half round its little harbour.   The chief part
of this day's journey, consists of a series of
long ascents and descents, the road winding along
the steep faces of the mountains almost overhanging the sea, at a great height, and every
now and then descending to the shore.

Splendid views along the coast, a succession of
bright blue bays and high precipitous headlands,
with glittering white towns, beautiful to look at
from a distance, seated on the projecting points,
and in the recesses of the shore.

Grand cliffs and singular masses of limestone.

Roccabruna, a little town stuck (as it were) against
the almost precipitous face of the mountain, rising
house above house.

Descend to Mentone, situated in an exceedingly
rich and beautiful plain of small extent, backed and
half surrounded by craggy mountains, peculiarly
wild and striking in their forms ; the sea in front.
Profusion of lemon and orange trees of very fine
growth, intermixed with olive trees of extraordinary
size.   The oleander planted in avenues.   The
Tamarisk forming large thickets on the sea-shore.
The little plain has an almost tropical richness
of vegetation, and the whole scene is very lovely.

Ruined castle on the hill above Mentone.

Vintimiglia, a picturesque old walled town, with
excessively steep and narrow streets ; a large modern
fort adjacent to it on the side towards Nice.

A most disagreeably steep and awkward descent

from the gate of Vintimiglia, to the torrent Roya,
which is crossed by a very long and abominably
narrow wooden bridge. Several other torrents,
some very broad, crossed between this and Bordig-
hiera.

It grew dark soon after we passed the Roya, and
we did not reach San Remo, our quarters for
the night, till eight o'clock.

From San Remo to Alassio. The coast, though
finely-indented is much less bold and beautiful than
that between Nice and Vintimiglia, owing to the
softer and more crumbling nature of the rocks; the
hills nearest to the sea, though steep are of moderate
height and tame outlines, and show no such towering
precipices or wild crags as we admired in our yesterday's journey; nor do any of the valleys appear to
rival in luxuriant richness those of Nice and
Mentone.

But the gay, white towns most picturesquely
situated, with their fantastic *campanili*, rising above
their clustering houses, form most striking features
in the scenery. Porto Maurizio is especially
picturesque. Nearly the whole of this part of the
coast, from San Remo to Alassio, is composed of
the formation of sandstone, shale and argillaceous
limestone, which I noticed yesterday as commencing
at Mentone. I believe it is the Macigno formation
of the Italian Geologists. There are frequent and
irregular alternations of its component members,

x

1848. but those which prevail most in these parts, are the
shale, which is very rotten and crumbly, and the
soft opaque, bluish-grey or pale brown marly limestone which resembles that used in lithography, and
has often a concretionary structure. There are
many transitions between these two substances, and
the shale also passes into the sandstone which is of a
dull brown or brownish-grey colour, sometimes micaceous and often more or less fissile. The strata of
these three rocks vary very much in thickness, and I
think in direction also, and are often much bent and
contorted. I met with no nummulites to-day,
indeed, in general all the three rocks seemed very
barren of fossils; but between Cervo and Andora, in
ascending a hill, I found in one place, some well
preserved fucoids (fucoides intricatus, or a closely
allied species) in the shaly beds.

These predominant rocks give a dull, grey-brown
colour to the exposed and prominent parts of the
coast, and as the shale is continually crumbling
away, and the more solid strata between which it
intervenes are thus undermined and fall down, the
steep declivities are everywhere covered with innumerable fragments and débris of the rocks.
There are no grand precipices like those formed by
the very hard and compact limestone of the Nice
coast, but the hills descend steeply to the sea,
generally very bare of vegetation, and cut by deep
and savage gullies. In this day's journey I did not
see a single plant of the Euphorbia dendroides, so
abundant on the limestone cliffs about Turbia and
that part of the coast.

The Oleander growing wild in the dry beds of two or three of the little torrents which furrow the sides of the hills between Oneglia and Alassio. Date palms cultivated in several places along this road.

Numerous old towers, low, massy and generally square, but now deserted and ruinous, stand on various points of this coast; built for protection against the Turkish and Moorish pirates.

Towns and villages very numerous, looking exceedingly pretty at a distance. Their streets are excessively narrow, so that it is difficult for a carriage to pass through.

January, 24th.

The inn at Alassio (la Bella Italia) is an old palace formerly belonging to the Durante family, and retains many traces of its original splendour. Between Alassio and Albenga, pass round a very bold headland; the road winds at a considerable height above the sea, and the descent below it is almost precipitous; yet the olive tree grows on this escarpment intermixed with the maritime Pine and the Caruba. This last is very abundant on the cliffs, and is a beautiful tree with its tortuous and knotted stem, and its rich, deep green, glossy foliage. The limestone of this headland appears (on a cursory view), much harder and less concretionary than that on the other side of Capo della Mele, and forms strongly-marked strata, without an intermixture of shale, so that the rocks altogether are much less

1848. crumbly.   Here the Euphorbia dendroides re-appears.

Albenga, situated in a broad and level valley, between two ranges of steep hills running down to the sea ; the eastern chain of hills exceedingly bare and brown. The valley richly cultivated, but wanting the almost tropical luxuriance of Mentone : —olives, vines, fig trees, corn, and groves of the Italian reed. Though the olive is still the chief object of cultivation, we do not observe it hereabouts of so great a size as about Nice, and in the more south-western part of the Riviera. A wooden bridge leading to Albenga, over the Centa, which is said to be one of the very few perennial streams between Nice and Genoa. We pass outside the walls of Albenga, the ancient Albium Ingaunum, a considerable town, fortified in the old style with battlemented walls, square towers, and a moat.

In the next headland, the rocks consist of a dark calcareous shale, full of veins of calcareous spar, which, as well as the laminæ of the rock, are twisted and crumpled in a most fantastic manner. Descent into another plain, in which are situated the towns of Ceriale, Borghetto, Loano, and La Pietra, inclosed by a fine amphitheatre of dark, rugged, craggy hills, behind which are seen higher mountains covered with snow. It was among those hills, I conceive, that the battle of Loano was fought in 1796, in which Massena defeated the Austrians.

Another cape, exceedingly bold and high, between this plain and Finale. Immense cliffs of very hard black limestone, through which the new road to

Finale is carried by a tunnel. The black limestone 1848.
is divided in a curious manner by joints in several
directions, so as to have something of a fragmentary
appearance. The old fortress of Finale, on the east
of the town on a precipitous rock, the sides of which
are half covered with the prickly Pear (Opuntia),
seemingly quite naturalized there.

Cape of Noli—magnificent cliffs of marble, towering from the sea, almost perpendicularly to a vast
height, and terminating in stupendous crags. The
sea below beautifully clear, of the purest and brightest aquamarine colour, lashing the fallen rocks and
fragments at the base of the precipice. The road
in this part is admirably made, cut in the solid rock,
and winding along the face of the cliff with a very
easy slope; in many places quite overhanging the
sea and passing through several tunnels. —
The limestone composing these glorious precipices
is very hard and compact, in some parts black, in
others reddish, and here and there green, everywhere
thickly veined with calcareous spar, and in fact a very
handsome marble. It is associated with a red and
bluish green shale. In some parts I observed a
beautiful breccia, of fragments of various coloured
marble embedded in a more crystalline calcareous
cement, this appears to form veins or fill up cavities
in the limestone. Anthyllis Barba Jovis grows
on these limestone cliffs. Euphorbia dendroides
very abundant, as also on the headland between
Finale and Loano.

Beautiful thickets of Mastic, mixed here and
there with Tamarisk, and interlaced by the twining

1848. Smilax, along the sandy shore about Loano and La
Pietra. The strap-shaped leaves of the Pancratium
maritinum appear above the sands of the sea-shore
in several places ; we first observed it near Alassio.
Euphorbia paralia abundant in the same situation.

I saw the Oleander growing wild in various places
among the rocks between Alassio and Albenga, but
only as a shrub—not a tree.

January 25th.

From Savona to Voltri, the coast continues of the
same general character—a succession of beautiful
bays, and high, bold prominent headlands, where the
ridges of hills come steeply down into the sea. In
the recesses of the bays, between these headlands,
occur in succession the towns of Albisola, Varazze
or Varaggio, Cogoleto and Arenzana. A great and
thick formation of stratified conglomerate, occasionally alternating with strata of red and greenish marl,
very similar in appearance to the upper new red
sandstone of England, begins a little on the Genoa
side of Albisola, and continues to Varazze. It is
evident that this conglomerate is of a later date than
either the black marble or the serpentine, since it is
full of rolled pieces of those rocks, some of the rolled
stones contained in it are of great size. The strata
appear to me to dip south-west, or thereabouts.
Immediately beyond Varazze, in the ascent of the
hill which projects into the sea between that town
and Cogoleto, the euphotide (diallage rock) shows
itself, constituting nearly the whole of that Cape ;
and strikes the eye at once by its difference in colour

and texture, and the forms of the rock-masses composed of it, from the formations we have hitherto seen along this coast. The compact felspar, which is its base or paste, is whitish; the diallage (bronzite) or a greenish silvery-grey, occasionally quite green, but rapidly becoming brown on the surface from exposure. The rock resembles the sienites of the Malvern Hills and of the Montagne de Tarare, in being divided by very numerous joints into irregularly angular portions. Though very tough and hard when fresh, it seems to disintegrate rapidly from the action of the weather; and hence it does not form mural precipices or projecting crags like the hard limestones, but constitutes rounded though very steep hills, deeply furrowed and torn by the action of water. It seems to pass into the serpentine, which makes its appearance a little before we reach Cogoleto, and continues to prevail with a few interruptions almost to the suburbs of Genoa. The serpentine rock here resembles in the form it assumes, and in its picturesque effect, that of the Lizard district in Cornwall. The wild irregular shattered masses at the foot of the promontories amidst which the clear bright green sea curls and foams, reminded me particularly of Kynance Cove. The varied colours, dark green and dark red and purple, relieved here and there by veins of white calcareous spar, are very striking.

In crossing the beautiful hill between Cogoleto and Aranzana, I observed near the top of the ridge a micaceous schist, resembling that which occurs below the lighthouse at the Lizard Point, and

1848. probably belonging to the same epoch as the serpentine. The blackish calcareous schist of Genoa appears in several places between Arenzana and the Polcevera, and it seemed to me as far as I could judge in passing hastily in a carriage, to alternate repeatedly with the serpentine. It is full of curiously contorted veins of calcareous spar.

The steep headlands between Savona and Voltri, and the chain of hills from which they jut out are clothed with pines, chiefly the Pinaster, the Pinus maritima, or Halepensis being here much less abundant.

From Voltri to Genoa is almost one continued town with innumerable gaily-painted villas and gardens rich in orange trees.

The splendid view of Genoa which opens upon one after passing the Lanterna, lost much of its effect through the gloom and haziness of the weather.

The road along this Riviera, from Nice to Genoa is much improved since I saw it before in December, 1827, and is in many parts a very fine road, constructed with great skill and labour and magnificence, broad and smooth and firm, and winding easily along the face of the precipitous cliffs. The portions from Nice to the valley of Mentone, and round the Cape between La Pietra and Finale and round the Cape of Noli, are particularly fine. In some parts the road is disagreeably narrow, and in many it would be improved by the addition of a parapet. The descent from Vintimiglia to the Roya is still very bad, and even dangerous.

The number of torrents to be forded is very great : 1848. most of them are at present nearly dry, having only a narrow and shallow stream winding through the midst of a broad bed of shingle, and some are entirely dry ; but after rain they swell very suddenly and often become for a time quite impassable. Several bridges, however, have been constructed quite recently.

This coast is certainly very beautiful. The alternations of high, bold rocky headlands with the graceful sweep of quiet bays and the contrast of steep pine-clad hills and towering cliffs and craggy mountains with fertile and richly-cultivated valleys, are delightful. The coast is exceedingly populous, and the multitudes of gay and picturesque white towns and villages which spangle it, contribute very much to the beauty of the scenery. One is much struck with the industry of the inhabitants, which can enable such numbers to subsist on such a narrow and rugged strip of coast. It is surprising to see the hill-sides cut into terraces and planted with olives and vines, in places where they are all but perpendicular, and to see cultivation in spots which appear almost inaccessible to human foot. Wherever there is cultivable ground along this coast, whether on the mountains or in the level valleys, the olive is the predominant object, and next to it the vine. Mulberry and fig trees grow to a great size ; and in the low grounds, the tall Italian reed is a constant and conspicuous and picturesque object in the landscape. I have noticed the groves of orange and lemon trees around Mentone ; and again be-

1848. tween Voltri and Genoa, the gardens are full of them. The Date Palm is cultivated in many places along this coast, beginning at Nice, but I have not seen it here of so fine a growth as at Naples and Palermo. Its young leaves are often tied up and bound round with matting, in order to blanch them, and they are afterwards cut off and carried in procession on Palm Sunday. The American Aloe appears to be half naturalized in some places.

The people of the Riviera are a picturesque and handsome race with rich brown complexions, very black hair and eyes, and very often good countenances. The red caps, red sashes, and brown capotes of the men are highly picturesque.

GENOA.

Pictures.—PALAZZO DURAZZO, STRADA BALBI.

1.—" Portia about to swallow the burning coals :" —by *Guido*, in his darkest manner. An exquisite picture : the face singularly beautiful, the expression admirable, the throat and bosom very beautifully painted. It is a picture to fall in love with—a face to dream of.

2.—" The Roman Charity," by *Guido*, in his pale style. Very graceful and pleasing : the daughter sweet and gentle, the expression of the old man's face and of his clasped hands natural and appropriate.

3. —" A Sleeping Child." *Guido*. Extremely pretty.

4.—" Magdalen." *Titian*. Very beautiful in colour ; a duplicate of the picture in the museum at Naples.

5.—" The Woman taken in Adultery." *Guilio* 1848. *Cesare Procaccino.* A fine and forcible picture ; but the head of Christ strikes me as ignoble and even vulgar ; the woman's head very fine.

6.—Portrait of a Bishop.—*Bernardo Strozzi.* *Il Cappuccino.* An admirable portrait. (Both Procaccino and Strozzi were Genoese).

7.—Portrait of a Little Boy in a blue dress, carrying a fish in his hand (in the character of Tobias). *Vandyck.* Quite charming.

8.—Portrait of a Little Boy, dressed in white satin, with his hand on the back of a chair. *Vandyck.* This also is charming.

9.—Portraits of Three Children together. *Vandyck.* A beautiful picture said to have been painted for Charles I. of England (representing his three sons) and bought on the dispersion of his pictures after his execution.

10.—Portrait of a Lady, seated, with two children. *Vandyck.* Beautiful. The children exquisite.

11.—Portrait of Philip IV of Spain. *Rubens.* A very finely painted portrait ; but of an unfavourable subject; the face indicative (very truly) of a mean character and narrow mind. Wilkie is said, when at Genoa, to have especially admired this picture.

12.—"St. Sebastian." *Domenichino.* Fine; but I am not sure that the attitude and expression are not too violent for a martyr.

13.—" Marriage of St. Catherine." *Paolo Veronese.* A small but fine specimen.

14.—" The Pharisees showing the Tribute-money to Christ." *Guercino.*

1848.   PALAZZO SPINOLA.   STRADA NUOVA.

1.—"Virgin and Child with St. Stephen and another Saint." *B. Luini.* Extremely beautiful, especially the head of the Virgin.

2.—"Dead Christ with the Virgin and another of the Maries" weeping over him. *Bernardo Strozzi. Il Cappuccino.* A very fine picture ; the calm but rigid lifelessness of the dead body, and the deep and strong grief of the two women, very forcibly given. It gives one a high idea of the artist, as does also his "Portrait of a Bishop," in the Durazzo Palace. But his genius was not suited to some classes of subjects, for his "Susanna and the Elders" in this same Spinola palace is abominable.

3.—"A Holy Family." *Guido.* In his "first" style. Beautiful, but I should never have guessed it to be Guido's.

4.—"A Holy Family," by *Mecherino* or *Beccafumi,* of Siena.

5.—A Fine Portrait of a man in a fur dress. *Titian.*

6.—"Dorinda Wounded with an Arrow by Silvio." (A subject from the "Pastor Fido,") by *Domenico Fiasella* of Sarzana. A pleasing and gracefully-treated picture ; the action natural, and the expression of the wounded woman very sweet and gentle.

7.—Luca Cambiasa painting his own portrait. A whimsical idea, but a very cleverly-painted picture.

PALAZZO   BRIGNOLE-SALE   (OR   PALAZZO   ROSSO) STRADA NUOVA.

1.—"Cleopatra." *Guercino.* Very fine. She is

represented lying on a bed, almost naked, with the asp still fixed on her breast, her eyes closed, her head sinking to one side, her hands drooping lifelessly: the whole attitude and countenance finely expressive of the calm sleep of death.

2.—" The Jews showing the Tribute Money to Christ." *Vandyck*. A picture admirable in colouring, and the heads of the two Jews very powerful; that of Christ less satisfactory. It is a picture much superior to that of the same subject by Guercino in the Durazzo palace.

3.—Portrait of the Marchesa Paolina Adorno Brignole-Sale. *Vandyck*. Admirable.

4.—Portrait of the Marchesa Anton Giulio Brignole-Sale, on horseback.—*Vandyck*. A most noble portrait.

5.—Portrait of another Marchesa Brignole-Sale, her daughter.—*Vandyck*. This also is a very beautiful picture, and there are two other excellent portraits by the same master.

6.—Portrait of a dignified and noble-looking Old Man, with a white beard and fur sleeves.—*Titian*. Very fine.

7.—Portrait of a Man with a black beard, in a black dress with crimson sleeves.—*Paris Bordone*. Very fine.

8.—" Virgin and Child," with various saints and little angels.—*Paris Bordone*.

9.—" Adoration of the Magi." *Bonifazio*.

10.—" St. Sebastian."—*Guido*. Beautiful.

11.—" Head of an Old Man praying before a crucifix."—*Luca d'Olanda*. Very singular.

GENOA.

Near the Palazzo Ducale two marble tablets
fixed in the wall, beneath the old tower, record the
names and crimes of two men—one Raphaelis de
Turri, the other of the family of Balbi, who were
doomed by decrees of the Genoese Senate to this
lasting record of their infamy, in addition to the
punishment of hanging, and to the confiscation of
their property, the destruction of their houses, and
the proscription of their families. The inscriptions
written in good Latin, enumerate a frightful catalogue
of crimes committed by them. This was certainly
an ingenious and formidable kind of punishment.—
The dates are 1650 and 1672. (M. Isola tells me
that the two persons thus denounced were political
offenders, who had conspired against the party then
in possession of the authority of the state).

PALAZZO PALLAVICINI, STRADA CARLO FELICE.

(Pictures).

1.—" The Magdalen carried to Heaven by
Angels."—*Franceschini.* Most graceful, sweet and
charming, somewhat in the style of Albani, but
superior to anything I have seen of *his,*—Franceschini, a painter who seems to be less known than he
deserves, was of the Bolognese School, a con-
temporary of Guido and Albani, and with much
similarity to them in his style.

2.—" Virgin and Child."—*Franceschini.* Very
sweet, resembling some of Guido's Madonnas, but
without the feebleness often apparent in his pale
manner.

3.—" Birth of Adonis."—*Franceschini.*

4.—" Diana surprised by Actæon."—*Albani.* A very good specimen of the master, the figures graceful and the attitudes spirited and expressive, but he has deviated singularly from the classical story, representing the goddess and her nymphs *not* naked, but in their shifts.

5.—" Coriolanus met by his wife and family in his advance on Rome."—*Vandyck.* This is really a group of portraits, put into the form of a historical composition, but the composition is bad, (independently of the anachronisms in costume) for the figures are all looking at the spectator, and not at one another, and have no appropriate expression. As portraits, they are finely painted, but not equal to some of those in the Brignole-Sale palace.

6.—" St. Francis Praying." *Guido.* (A small picture and half-length in the same room with the Albani). The expression of earnest devotion seems to me fine. There is a large picture of the same subject in another room also ascribed to Guido, but I do not like it so much.

7.—" St. Jerome." *Guercino.* Genoa. Palace of Prince Doria. This great palace, the residence of the illustrious Andrea Doria in his latter years, is historically the most interesting in Genoa, and is one of the finest, perhaps the very finest, in point of situation. The building itself is not handsome, but its great extent of front renders it very conspicuous. Its gardens with their marble balustrades and hedges of clipped evergreens, their grand fountains designed by Carlone, and their terraces overlooking

1848. the sea, are noble, though in a state of great neglect; and the view from them of the city and bay of Genoa is most beautiful. In the garden above the palace, on the slope of the hill, is a colossal statue of Jupiter, conspicuous a long way off; and beneath this we are told is the tomb of a great dog which was given to Andrea Doria by the Emperor Charles V.

The paintings on the ceiling of the entrance hall, representing the triumphs of Scipio, and various stories from Roman history together with a variety of mythological and allegorical figures, were executed by Perino del Vaga (Pietro Buonaccorsi), a scholar of Raphael, who having fled from Rome when it was sacked by the army of the Constable de Bourbon, took refuge at Genoa, and was patronized by the great Andrea. He painted also the beautiful arabesques on the ceiling of the staircase, the series of figures of heroes of the Doria family on the walls of the gallery, the groups of Cupids or *putti* above them, and the historical or mythological subjects on the ceiling of the same. The putti are of extraordinary beauty, I scarcely remember to have seen anything equal to them in their way. The ceilings of many other rooms are covered, some with fresco paintings, some with bas-reliefs in stucco, representing various mythological stories, intermixed with arabesques in the style of those in the Loggie of the Vatican, of amazing variety, delicacy and beauty. These frescoes also were painted by Perino del Vaga, but have been re-touched or re-painted in modern times. There are also

some fine bas-reliefs in marble, in medallions by 1848.
Montorsoli.

GENOA.—PIAZZA DI S. MATTEO.

This little piazza is an interesting specimen of *old*
Genoa. One side of it is formed by the curious old
Church of St. Matteo, with its façade of black and
white marble in alternate courses, and with long
inscriptions in old characters on the white marble,
commemorating the good deeds of the Dorias, for
this Church was founded and endowed and kept up
by the family. It stands on a platform, approached
by flights of steps from the opposite side of the
piazza. On another side is a fine old house, very
lofty and built of black and white marble like
the Church, and of great interest as having been
given by the Senate of Genoa to Andrea Doria,
—"A public gift to the liberator of his country"—
as is still recorded by the inscription over the door-
way. This house is now occupied by the Consulate
of New Grenada. The marble door-posts and
lintel are sculptured with very elegant arabesques.
On the remaining sides of the square are other tall
old houses, black and white ; and the whole has
a venerable and dignified aspect, quite in character
with the old aristocracy of Genoa.

LETTERS.

My Dear Father,

1848.
        I thank you very much for your letter which
I found at the bankers on our arrival here, and
I feel very much your kindness in exerting yourself
to write to me while you were so much oppressed by
the effects of the influenza.   I am exceedingly sorry
that you continued to suffer so much from that most
disagreeable and depressing malady, and I fear the
climate of Suffolk is but very ill calculated to
promote your recovery from it.   I wish you might
be induced to spend next winter either at Nice or
Rome, which I cannot help thinking would be much
better for you.

We arrived here three days ago, but we have
come too soon, for Genova la Superba is most
abominably cold,—very nearly as bitter as Barton or
Mildenhall.   A great deal of snow fell last night
and this morning, and the streets and balconies are
all white with it, nor does it seem in the least hurry
to melt ; and this is the third heavy fall of snow
they have had this winter.   The wind is from the
north, very strong and horridly cold.   We cannot
help (at least Fanny and I) looking back with some
regret to the delicious weather we enjoyed at Nice,

and wishing we had stayed there longer. Fanny and Susan however are much pleased with what they have hitherto seen of Genoa, which is certainly a far more interesting place in itself than Nice; and we are most comfortably lodged in the Hotel Feder, where the people are beyond anything attentive and obliging.—I called yesterday on Mrs. Granet, and on Isola, both of whom received me with the utmost cordiality, and seem disposed to do everything to render our stay agreeable. Both enquired much about you. Mrs. Granet gave me an interesting account of the *demonstration* which took place here on the 10th of last month, and which she witnessed. Twenty-four thousand persons, of all ranks, among them Madame Doria and several other noble ladies, walked in procession to the spot where a stone in the pavement commemorates the breaking out of the insurrection against the Austrians in 1746. Among all these thousands there was no disorder, no rude pushing or jostling, no quarrelling, nothing (Mrs. Granet says) disagreeable or alarming to the most nervous person. A certain number of gentlemen, who had suggested and arranged the procession, had told the Governor that they would engage to keep order if he would keep the police and soldiers out of the way; accordingly the troops were kept within the barracks and forts, and a certain number of the police in plain clothes were placed under the direction of these gentlemen to support them in case of need, but with strict orders not to act without *their* authority. During that day and night, nearly eighty-thousand persons were

1848. collected about that part of the town, and continued
all night waving banners and singing patriotic songs,
but there was no disorder, no robberies, no violence,
no insults to any one ; the people shouted for those
whose sentiments and conduct they approved, but
offered no insult to those who were known to be
of the contrary party. Once, indeed, some of them
proposed to go and shout, "a basso i Gesuiti!"
before the Jesuit's college, but they were easily
dissuaded from this by the gentlemen who had
influence among them. The whole thing was most
honourable to the Genoese character.

The excitement however went on increasing, the
nocturnal assemblages were so frequent and noisy as
to create some alarm, and at last threats of violence
began to be heard towards the Jesuits and other
unpopular personages, so that the Governor thought
it necessary to interfere, and the proclamation of the
3rd of this month, was issued for bidding the nocturnal assemblages and the shouting and singing in
the streets, and in short calling the Genoese *to order.*
That day and the next, Mrs. Granet says, would
have been anyone who had not the confidence in the
Genoese which she has, for everything was prepared
by the Government to use force in the most effectual
manner if necessary, but all passed off quietly.—At
present, the town is as free from anything like disturbance or tumult as can be, but it is very interesting to observe the symptoms of the excitement
which pervades men's minds, though silent and
tranquil,—the booksellers' shops all full of patriotic
pamphlets and songs and political publications, some

original, some translated from English or French ;
Liberal newspapers established; and what strikes me
as a remarkable *local* sign of the times, at every turn
one sees rough prints and popular narratives of the
expulsion of the Austrians in 1746. I like very much
the tone and spirit of those of the new patriotic publications which I have yet seen, they are temperate and
rational, and manly without the least taint of
Jacobinism, and show a respect for the rights of all,
and a high moral feeling. God grant that the
Italians may go on as they have begun, and not
sully their good cause by any excesses.—I forgot to
tell you that a few days ago there was a solemn
funeral service, celebrated here in San Siro, to the
memory of those Liberals who were killed in the
commotions at Pavia a short time since. We
enjoyed very much our stay at Nice : the weather
was delightful, the place very pretty, our excursions
in the neighbourhood very pleasant, and the society
of the Napiers was a great pleasure to us, so that
Fanny and I were very sorry to come away ; but we
hoped to find letters here (in which we have not
been disappointed) and Susan longed to be really in
Italy, and to see pictures, of which there are
none at Nice. We hired a vetturino, left Nice on
the 22nd, and came hither by the Corniche, a
very interesting journey,—but we committed the
error of travelling too fast, for we came in four days,
which at this time of year when the daylight is too
short, does not allow sufficient time for seeing.
The English generally make the journey in three
days. We slept at San Remo, Alassio and Savona,

1848. baiting in the middle of the day at Mentone, Oneglia, Finale and Voltri. The coast is certainly very beautiful, the succession of headlands and bays, and the contrast of rugged hills and bold precipices and snowy mountains with the rich cultivation and smiling beauty of the valleys produce an extraordinary variety, and afford a continual delight to one's eyes. I was particularly struck with the grand scenery about Esa and Turbia, the charming valley of Mentone, and the magnificent marble cliffs of the Cape of Noli. The road is very much improved since I travelled it before with you in 1827, and is now on the whole a very fine road, and in many parts admirably constructed, but it would often be much improved by the addition of a parapet, and the descent from Vintimiglia to the bridge over the Roya is still frightful. But we were most near meeting with a second overturn almost at the very gates of Genoa, between Sestri and San Pier d'Arena, where from some cause or other, the road was in a most abominable condition. Unfortunately, the day we arrived here was very hazy and gloomy as well as cold, so that the splendid view of Genoa which opens upon one as soon as one has passed the Lanterna, lost much of its effect.

Mr. Abercromby is at Turin, and Lady Mary is still with Lady Minto, but is expected to arrive here very soon, on her way to rejoin her husband.

Pray give my love to Emily, and my thanks for her letter which I received at Nice.

Believe me ever, your very affectionate son,

C. J. F. BUNBURY.

My Dear Emily,

We have been eight days at Genoa, and this is the first pleasant day we have experienced; we had bitter cold winds, as cutting as you enjoy at Barton, then snow, which lay for two days even in the streets, then heavy rain, and then another little fall of snow; till I began to be quite in despair; but to-day has been really fine, bright and sunny and mild, like the weather we enjoyed at Nice, and we have at last seen the superb city to advantage.

This is a time of great political excitement and interest, and one cannot but sympathize with the enthusiasm of the Italians for their newly revived freedom, and their bright hopes of amelioration and progress. Fanny, who you know is not of a cold, or unexcitable disposition, is wild about Italian politics, and thinks and dreams of nothing else. We have just made acquaintance (by means of a letter from her friends at Paris) with Count Mamiani* a distinguished Liberal, for a long time an exile, but now residing here (although he is a Roman) and an active contributor to the *Lega Italiana*, a Liberal newspaper, very lately established here. His conversation is interesting, and like what I have read of his writings, shows much ability, and an excellent spirit, moderate, humane, thoughtful and considerate. Our friend Isola also is a zealous Liberal; and from all I can hear, there seems to be a remarkable unity of feeling throughout the people of this country, and

* Afterwards Prime Minister at Pio Nono.

1848. a most warm and hearty sympathy with the
Liberals in all other parts of Italy. In my letter of
the 28th, I told my father what I had heard from
Mrs. Granet about the proceedings here on the 10th
of December and since, and the indications we had
ourselves observed of the popular feeling. Since
then a fresh excitement has been created by the
news from Naples. Early in the morning of Sunday
the 30th, a Neapolitan steamer of war arrived in this
port, and the Neapolitan Consul immediately went
on board, but nothing was officially divulged as to
the news it brought; rumours however soon spread
that either the Minister Del Carreto, or some said
even the King of Naples himself, was on board—a
fugitive. The Captain would not allow anyone to
go on board, and the crew would not answer any
questions. A great crowd assembled before the
Neapolitan Consul's house, uttering furious cries
against Del Carreto, whom they supposed to be concealed there, and it was with difficulty that the
Consul pacified them by assurances that the odious
Minister was not in his house, and had not landed.
The steamer departed immediately after taking
in supplies of water and coals, having arrived in
such an unprovided state as to show plainly that
she must have left Naples very suddenly, and at
very short notice. In the afternoon there were fresh
assemblages and outcries before the Neapolitan
Consulate, the people expressing loudly their sympathy with the Sicilian insurgents, and their
detestation of Del Carreto; and then the Consul
acknowledged that Del Carreto *was* on board the

steamer which had just left the port, but declared 1848. that he was gone on in it for Marseilles. And this is believed to be true. It is asserted also that the wretch was obliged to leave Naples in such haste that he had not time even to dress himself. Yesterday morning the news arrived that the King of Naples had granted a Constitution, and had re-called his troops from Sicily, and the enthusiasm here was great. The principal streets were full of people all day long, and great numbers wore ribbons of the Italian tricolor—red, white and green ; and at night there was a general illumination, which had a very beautiful effect, especially in the Strada Nuova. Susan and I walked, and Fanny went in a Sedan chair, through all the principal streets and piazzas, to see the illumination ; there were great multitudes of people, very enthusiastic, shouting and singing their patriotic songs, but all in high good humour, and perfectly orderly, sober and peaceful ; no disagreeable crowding, no shoving, no disturbance or outrage of any kind. I never saw people behave better. I must say that from all I can see and hear, no people appear to me more worthy of freedom than the Genoese. It is delightful to see the political re-generation of such a country as Italy. One of the best symptoms is the unity of feeling which seems to pervade all parts of the country, the sympathy between the various populations and the desire to co-operate and aid one another, and to lay aside and forget as far as possible the old divisions and enmities between different parts of Italy, The patriots here are vexed at the conduct of the Sicil-

1848   lians in striving for a " repeal of the union " between their own island and Naples; but it seems to me very natural in a people so abominably oppressed as the Sicillians have been. The people of Palermo are said to have fought very hard, and certainly their success appears very surprising, unless we suppose (which is just possible) that the Neapolitan troops did *not* fight very well. We hear nothing of the killed or wounded. It is rumoured that the Austrians intend to seize upon Alexandria—if they can ; but I hope the Piedmontese and their king will not be caught napping.

(February 6th). We were agreeably surprised yesterday morning by the sudden arrival of the William Napiers at this hotel, and on enquiry we found that Emily's health had been so delicate that she had been advised to leave Nice, the air of which was too sharp and exciting for her chest, and to spend the remainder of the winter and the early spring at Pisa. I trust that there is as yet nothing seriously the matter with her, and that by this timely precaution she may escape all danger. They went on yesterday evening by the steamer to Leghorn. It will be a great pleasure to us to see them again at Pisa, where we propose to spend a week in the same hotel with them, and if they spend the summer at Castellamare, as they talk of doing, we may very likely be companions there also.

We went to the theatre last night, Mrs. Granet having lent us a box, and between the 2nd and 3rd acts of the opera (which was the Horatii and Curiatii) there was an amusing interlude, the pit called loudly

for the National Hymn, and made a great clamour, 1848.
till the whole corps dramatique came on the stage
adorned with cockades, and waving banners of
the favourite colours (the red cross of Genoa, and
the Italian tricolor) sung in chorus the new patriotic
hymn, which had a very fine effect, the applause was
great, and the enthusiasm so contagious that I
joined with all my heart in it   Indeed this was to
me by far the most interesting part of the perform-
ances.—I am very much obliged to my father for his
letter of the 27th of January, which I received on my
39th birthday, and I am very glad to learn from it
that you were so much better. Isola has been in
the highest degree obliging and useful to us, and we
have seen many palaces and galleries under his
guidance. He begged me to remember him most
warmly to my father, and to thank you very much
for a book which you sent him. I have hardly room
left to tell you how much we enjoyed our stay at
that charming place Nice, and how sorry we were to
leave it, or how many pleasant excursions we made.
I shall always retain a very agreeable recollection of
Nice, and as for the climate, I should call it one of
the most delightful winter climates I ever experienced, equalled only by that of the Cape. I was
glad to find by the way, that my little book* gave so
much gratification to your brother, Sir George.

We propose, weather and other circumstances
permitting, to leave this place on Friday next, the
11th, and halting a day at La Spezzia, to reach Pisa
on the 15th. Fanny sends her love. I read in

On the Cape of Good Hope.

1848. *Galignana* the sad account of the massacre of five
unfortunate officers by the Caffirs, but I did not
then know that one of them was one of your cousins
the Bakers.

I am very sorry for his family.

Believe me ever,

Your very affectionate step-son,

C. J. F. BUNBURY.

<hr />

Genoa,
February 13th, 1848.

My dear Father,

We shall set off for Pisa to-morrow morning,
but before quitting Genoa I will give you some
account of what has happened since I wrote.

My last letter was to Emily, giving her an account
of the rejoicings here on account of the success
of the Neapolitans ; but the excitement was far
greater on Wednesday the 9th, when the King's
proclamation arrived from Turin, establishing a
representative and constitutional government for all
his subjects, with two Chambers, responsibility of
Ministers, &c. &c. The enthusiasm was beyond
anything I ever saw : all Genoa seemed to be in the
streets ; thousands and tens of thousands of people
of all sorts and classes ; men, women, and children,
nobles and mechanics, soldiers and citizens, and
even priests, walked in procession round the town
with banners, and singing the patriotic hymns ;
every one wore ribbons or scarfs of the King's
colour (dark blue) very often combined with the red
and white of Genoa. At one o'clock there was

a solemn thanksgiving service in the Cathedral,— we did not go in, but it was a fine sight to see the steps and the whole space before the venerable old building thronged with people, with banners innumerable ; and to see when the benediction was given and the people knelt, the banners all sink down at once, as if levelled by a mighty wind ; and then rise again when the ceremony was finished.

All day long the stir in the streets and the processions, and the singing continued, and all faces were full of joy and exultation. Our friend Isola was quite radiant with joy : "à present nous sommes hommes," he exclaimed when he first met us. But it was droll to see bands of little children marching in procession, and chanting away with all their might. They could at least understand the pleasure of a holiday, and of making a noise. At night every house was illuminated from top to bottom, and the streets were still more thronged than in the day. The processions then carried torches, and the effect was beautiful when one looked down on them from the windows and saw the endless column of blazing lights, winding slowly through the narrow streets. We walked about the town that evening till we were tired, we three and Mr. Isola, and found the crowd and the noise tremendous, but with all the excitement, there was nothing like tumult or disorder, no appearance of outrage or violence, no quarrelling or fighting, no robbery, no angry cries. It is impossible for any people to behave better than the Genoese have done through all this stirring time.

(February 19. *Pisa.)* We arrived here the day

1848. before yesterday, having had a delightful journey of five days from Genoa, sleeping at Sestri, Borghetto, La Spezzia, and Pietra Santa. The scenery all the way from Genoa to Sestri di Levante is quite charming, — we preferred it to the other Riviera, and we had delicious weather to give us the full enjoyment of it. But above all we were delighted with Sestri : it is one of the loveliest spots I ever set eyes on, a perfect gem, a retreat fit for Calypso. If the political disturbances should prevent our going to Naples this year, I should much like, and so I think would Fanny, to spend the summer at Sestri. La Spezzia too is delightful : we arrived there at noon, and so had time to enjoy it. The mountain scenery between Sestri and Borghetto is of a wild and singular character, and not without grandeur in its savage rudeness, and we were very much struck with the view of the magnificent Carrara mountains, which opened on us when we reached the highest point of the pass. Those "Alpes Apuanæ" are by far the finest of the Appenines that I have ever seen ; they made a glorious show again when we were crossing the Magra. I wished to have made a halt at Carrara, but was deterred by the accounts which we found recorded in all the travellers' books and confirmed by our vetturino of the intolerable badness of the inn there ; it is a pity there should not be a decent hotel at such an interesting place, for besides the beauty of the scenery and the geological interest of the marble mountains, the cathedral is exceedingly curious. The beautiful cultivation of Tuscany

and of the plain below Massa lost much of its 1848.
effect from the season, the trees and the vines
being still leafless; indeed I think I have hardly yet
seen so much appearance of Spring as we found
at Nice in the beginning of January; but yet
our tour was very enjoyable. We found very nice
inns at all the places where we slept between Genoa
and this place. We are all charmed with Pisa,
as far as we have yet had experience of it, and
especially with the group of glorious old buildings,
the cathedral and its companions standing so
majestically apart, "fortunate alike in their society
and their solitude," and free from any modern
incongruities that can disturb the images of ancient
days which they impress on the mind. Even the
beggars, though they often disturb one's contempla-
tions, are not incongruous; for in Andre Orcagna's
great fresco in the Campo Santo, there is a group
of beggars evidently drawn from the life, the very
patterns of those that pester one at the present day :
faces, attitudes, costume, everything has been preserved unchanged among the mendicant fraternity
since the fourteenth century. Susan is delighted
with the Campo Santo, where she spent the greater
part of yesterday, and where she has got leave
to draw.

The William Napiers are established in very
quiet and comfortable lodgings, very near the
cathedral, and we shall see them, I hope, constantly
during our stay here, for both Fanny and I like
them very much. William has just gone off by the
railway to Leghorn, in the expectation of meeting

1848. Captain and Mrs. Mac Murdo, who are on their way
to Nice.

To Fanny's great satisfaction we found plenty of
political excitement all the way as we came along
except (of course) in the territory of Massa Carrara.
The evening we arrived at Sestri, that little town,
and Chiavari and Lavagna, and all the towns along
the Riviera, were illuminated in honour of the
Constitution; at Spezzia we heard of similar rejoicings, and the morning after our arrival at Pietra
Santra, there were great festivities *there*, for the
Constitution which had just been promulgated by
the Grand Duke. We were too late (much to my
satisfaction) for the principal rejoicings here, but we
found almost every house on the Lung' Arno,
and in some of the other streets, adorned with
flags of the favourite colours, which had a very
gay and pretty effect. So far all is well. But
one cannot help feeling great anxiety about the
turn which events may take either in the south or
north. It is reported, as I see, in Galignani, that
the Sicilians had agreed to accept the Constitution
offered to them by the King, *on condition* that it should.
be guaranteed by England. The state of Lombardy
seems to be terrible; the Austrians are tyrannising
more than ever, and Radetzky is said to be going on
as if he wished to drive the people to desperation.
Most naturally a vehement sympathy is felt for
the Milanese by those Italians who are already
fortunate enough to enjoy political freedom; and
here, as it seems to me, is the great danger for
Italy : if the King of Sardinia should be hurried by

the impatience of his subjects into an attempt to drive the Austrians out of Lombardy, I fear that these latter might be able, not only to hold their own, but to recover by force of arms, the ascendancy which they have now in a great measure lost over the rest of Italy. I heartily wish that this danger may be averted, and that time may be allowed for the good works of the Pope, the Grand Duke, and the King of Sardinia, to work out their natural and wholesome effect; as I am convinced that the effects of constitutional freedom in these parts of Italy will be most salutary, if not nullified by war or foreign interference.

The weather here is very changeable : yesterday was beautifully fine, to-day it is raining cats and dogs. I cannot tell exactly how long we shall stay here, but very possibly a fortnight. I trust that Cecilia is quite recovered from her attack, and is enjoying Torquay, and that you have perfectly got over the relics of the influenza. I cannot tell you how kind and obliging and attentive Isola was to us while at Genoa, or how he exerted himself to do the honours of the place, and to show us everything worth seeing. We all liked him very much.

Believe me,

Ever your very affectionate Son,

C. J. F. Bunbury.

                                              Florence,
                                       March 12th, 1848.

My dear Father,

          We arrived here four days ago, and have
been extremely busy till to-day in seeing sights, and
now I take advantage of a wet Sunday to write to
you. My last letter to you (begun at Genoa and
finished at Pisa) contained such an account as I
could give you of the proclamation and reception of
the Constitutions of Sardinia and Tuscany ; but
since then the importance of Italian politics has
been quite swallowed up in that of the astonishing and
really awful events in France. I do not think I ever
heard of a revolution of such tremendous extent and
importance effected with such incomprehensible
ease and speed. It is quite bewildering and hardly
seems like reality. Who could have dreamed that
Louis Philippe, with all his armies and fortifications
and all his sagacity, all his experience, all his selfish
cunning, would fall even more easily than Charles
the Tenth, and in his old age be driven a second
time into exile ! It is a more wonderful illustration
of the " vanity of human wishes " than anything in
Juvenal or in Johnson. I cannot much pity the old
man, whose whole course of policy for many years
past has been pre-eminently selfish, still less am I
sorry for Guizot, who has spent the years of his
political power in a constant practical opposition
to all the noble principles which he had advocated
in his writings. But I cannot help feeling anxious
and somewhat nervous about the future. There is
something awful even in the idea of a French

Republic, from the memories with which it is 1848. associated, and though they have begun well so far, and made the fairest professions of peace, some of the names included in the new Provisional Government are by no means such as to inspire confidence. The projects they have announced, and the engagements they have entered into with respect to the working classes, look as if they were under the influence either of very extraordinary theories, or of a dangerous " pressure from without." I heartily wish that things may go on well in France,—but time must show, for I hardly think that the wisest and most experienced politician would venture to prophesy what may come to pass in that country within the next two years. At any rate I trust and believe that the English, both government and people are now wise enough not to interfere with any form of government that their neighbours may choose to adopt, and to remain at peace with France so long as she is willing to be at peace with us.—The Italians very naturally rejoice in the change, for they are sure of sympathy and support from a French Republic, whereas they were well aware that Louis Philippe would do all he could underhand to thwart and injure the cause of liberty. I hope they may not be stimulated by the success of the Parisians to rash haste or impatience. I have nothing particular to tell you about Italian politics : all is quiet here, and strangers as safe as ever they were. The Civic Guard is *the rage* in Tuscany, every second or third man you meet wears the military cap, and you see little boys carrying miniature muskets and rehearsing

1848. the movements of the parade. I do not know whether the English newspapers may have mentioned some interesting proofs of tolerance and true liberality of spirit which Italy has lately afforded. The new Constitution of Tuscany proclaims the civil and political equality of all Tuscan subjects *without regard to their religious opinions*; in Piedmont, the King has emancipated the Waldenses, relieving them from all the restrictions and disabilities to which they had been subjected by his predecessors, and amidst the rejoicings for the Constitution at Turin there were loud cheers from the people " for their Jewish brethren,"—and a deputation was sent to compliment the pastor of the Vandois, and to express the good will and friendly feeling entertained by those of the established Church. I am afraid that England is much behind-hand in religious toleration. We have seen Signor Gaetano Turchi, and the beautiful drawings* (very beautiful they really are) which he has made for you. He is waiting for your instructions as to the mode of sending them to you.

We are in ecstacies with the pictures, statues, and buildings here, and are making the best of our time in seeing the sights. Fanny is so much delighted, and so eager, that I have great difficulty in restraining her from wearing herself quite out. We are extremely well situated in the Hotel de York, looking on the cathedral and its beautiful campanile, which Fanny is at this moment engaged in drawing. We have been twice to the gallery of the *Uffizii* and once to the Pitti, and have seen the cathedral, Santa Maria Novella, and S. Lorenzo. Susan has

* Copies in sepia of Andrea del Sarto's frescoes in the Scalzo.

fallen in love with Fra Angelico da Fiesole, and has 1848. begun copying one of his pictures in the gallery: and I, less angelic in my tastes, have fallen in love with Titian's " Flora." But though I cannot go so far as many persons do in the present day in admiration of the very early masters, I certainly appreciate their merits much more than I did before this visit to Italy. The frescoes of Orcagna Pietro Laurati, and Benozzo Gozzoli, in the Campo Santo at Pisa and those of Simone Memmi in Santa Maria Novella, have given me a high idea of their power, variety and richness of invention, which I confess 1 should not have estimated so highly from their oil pictures, or rather easel pictures. I have not yet arrived at a proper comprehension of the merits of the " Beato An- gelico," but perhaps I shall in time.

Do you know Lord Lindsay's book on "Christian Art ?" It is a very interesting and a most useful guide to the study of the early painters. Among the pictures here which have particularly delighted Fanny and me (besides Titians, never-to-be-for- gotten "Flora") are the " Madonna del Cardellino," Pietro Perugino's " Deposition from the Cross," Titian's "Magdalen," sundry portraits by the same, especially those of Pietro Aretino, Luigi Cornaro, Giovanni de Medici, and a Lady without a Name, in the first room of the Pitti palace. The "Ma- donna della Seggiola" (or course) and one picture of Andrea del Sarto, in which four saints, standing, are the principle objects. But one is almost bewildered among such a variety of admirable things as are to

1848. be seen here. Florence pleases me even as much as I had expected, and I shall leave it with extreme regret, but we must not stay here much longer, or we shall be too late at Rome. We talk of setting off to-morrow week, and stopping a day at Arezzo and two days at Perugia.

We liked Pisa very much, and the society of the William Napiers added much to the agreeableness of our stay there,—indeed I do not think we should have remained so long but for them. Susan however found plenty of employment in the Campo Santo. Major and Mrs. MacMurdo arrived while we were there, but stayed only a few days and then went on to Nice; we liked much what we saw of them. During our stay at Pisa we made excursions to Leghorn and Lucca, in each case going by the railway and returning the same day. At Leghorn we did not see anything interesting except the English cemetery, but at Lucca we admired the cathedral, the pictures of Fra Bartolommeo there, and in S. Romano, and one of Francia in S. Frediano. Lucca is an interesting city, and Fanny was much delighted with its many curious old churches, but was very indignant that there should be no satisfactory engravings of them. On the whole, the Italians seem to care very little for the architecture of the middle ages.

Fanny sends her love and so does Susan; and I beg you to give mine to Emily and Cecilia. I hope to hear before very long that you are all well.

Believe me ever, your very affectionate Son,

C. J. F. BUNBURY.

Rome,
April 6th, 1848.

My Dear Father,

On our arrival here, I had the pleasure of
receiving your letter of the 3rd of last month, which
was very interesting to me, and for which I thank
you much. My last letter to you was from
Florence, and was written before the effects of the
French Revolution on Italy had developed themselves. We live indeed, in stirring times, and
events succeed one another so rapidly that it is
difficult to keep pace with them. Anyone could
see that the extraordinary change in France must
have a very powerful influence on the course of
events in Italy, but I certainly had no idea that the
explosion in Lombardy would have been accelerated
and facilitated by a Revolution at Vienna, the very
last city in Europe where I should have thought a
Revolution likely. I had always imagined that, in
Austria itself, the Emperor and Metternich were
looked upon as infallible, and that popular ideas
were as little likely to disturb the peace of Vienna
as of Pekin ; but I find I was quite mistaken. We
were at Florence when the news of this affair
arrived (with the usual exaggerations) and of course
the stir was great, but when a day or two after there
came the news that Lombardy was up in arms, that
the people had triumphed at Venice, and that
Modena was revolutionized, and that fighting had
begun at Pavia, Milan, &c., then the ferment was at
its height. Thousands of people thronged the
Piazza del Gran Duca and the Loggia dei Lanzi,
calling out for arms, and demanding to be led

1848. against the Austrians. The Grand Duke, taking his course with promptitude and spirit, immediately issued a proclamation, announcing that the time had come to strike a blow for the regeneration of Italy,—that he had ordered all his regular troops to march at once for Milan ; that any volunteers who wished to follow them should be placed under experienced officers, and that he entrusted the defence of himself and the city to the civic guard. This was the day before we left Florence. The troops were to march the next day, and at every turn one saw volunteers preparing their equipments, or hurrying to the points of rendezvous. Still it struck me that amidst all their eagerness, the people were much quieter and less noisy than at Genoa. When we entered Umbria, we found that the Pope had likewise put his troops in motion ; between Foligno and Terni, we met great numbers on the march, and at Terni we saw a corps of about 1500 regular infantry, and a considerable body of volunteers, march into the town and parade in the piazza, a very animating sight. All along the road from thence we continued to meet frequent detachments, chiefly of volunteers, who seemed full of eagerness, and cheered lustily as they passed us, but it seemed more like the eagerness of school boys going on an expedition which promised " lots of fun," than anything more serious. The inns at Civita Castellana and other places along that line were crammed with soldiers, and it was with difficulty that we got accommodation for the night at Le Sette Vene.

Since our arrival here, we have learnt that this

volunteering was encouraged from motives of policy 1848.
by the Papal Government, which induced some
thousands of the most idle and disorderly and
dangerous characters in Rome to enrol themselves
for this expedition against the Austrians, in order to
keep them out of mischief ; and the city is said to
have been much quieter since their departure.
Gibson (the sculptor) says that when the volunteers
were mustered for departure, and a great multitude
of people assembled to see them march, he heard a
girl, in taking leave of her lover, beg him to bring
her back an Austrian's head as a present!
This sentiment was much applauded by the by-
standers.

A great many English have taken fright, and
have left or are leaving Rome, among the
number, Cecilia's Cape friend, Sir Anthony
Oliphant. I do not believe that there is any
danger, at least in Rome itself, though perhaps
the country may be less secure than it was ;
but it is true that one must trust to the good nature
of the people, for the Government seems to have
little or no power. They are a singular people
these Romans ; Gibson tells us that when the
tidings of the Revolution of Vienna arived here, a
vast multitude assembled before the palace of the
Austrian Ambassador (the Palazzo di Venezia) and
sent up a deputation of twelve men, who entering
the Ambassador's room, with the utmost politeness,
requested him to take down the Austrian arms from
the front of the palace ; he declined ; upon which
the deputation retired without saying anything

1848. more ; and the mob, having with much trouble succeeded in obtaining a ladder sufficiently long took down the coat of arms, tied it to the tail of an ass, dragged it all along the Corso to the Piazza del Popolo, and there burnt it. They then went to the Austrian Church, took down the arms from thence, and beat them to pieces. And in this work of destruction they were occupied for at least half a day, without interruption from military or police, and without hurting anyone, or injuring any private property.

Gibson says also that in the midst of all the popular vehemence against the " Tedeschi " (for they do not distinguish between the Austrians and the rest of the Germans), not one of the many German artists residing at Rome has been in the least injured or even annoyed. It is to be hoped however that the volunteers who have marched to the north will not come back for some considerable time, as Rome would not be a very pleasant or safe residence when they returned. I am very anxious to learn how the news of the insurrection of Lombardy is received in England and France. I think the French are not at all unlikely to take an active part in support of the Italians, for as the Government of that country seems to be daily plunging deeper and deeper into financial and economical difficulties and disasters, and getting more and more inextricably entangled by their rash engagement with " the masses "—they may be very glad to cut the knot by plunging into a popular war. I do not suppose the Austrians, in spite of their own

Revolution, are likely to give up their hold on Italy 1848. without a blow ; indeed, I wish they may not, for though I heartily wish the Italian patriots may be successful in the end, it would do them a great deal of good to have a tough struggle for it ; this would sober them, and take down their conceit, which at present is excessive. If the French interfere on behalf of the Lombards, Russia will no doubt step in to support Austria, and then all the Continent will be in a blaze. And will England be able to keep clear ?

I am very happy to find by all accounts that amidst this general turmoil and crash of thrones and old institutions, things remain so quiet in England. The state of affairs in France looks to me very bad ; the power is falling more and more completely into the hands of the classes who have no property, and those not of France in general, but of Paris—in fact, as it seems, into the hands of the *classes dangereuses* of Paris. The National Guards who really made the last Revolution, have allowed themselves to be thoroughly bullied by the mob, and though, as yet, no sanguinary tendency has shown itself, the Parisians seem ready to run, in other respects into extravagancies as wild as those of the first Revolution. How will it all end ? I should not be at all surprised if, before another year, the members of the Provisional Government (or at any rate the best of them) became exiles in their turn.

Since we came here we have had a good view of the Pope, at a very grand procession in the Piazza

1848. of St. Peter's. The occasion was that a most
valuable relic, the head of St. Andrew, which had
been stolen sometime ago, was recovered, and was
brought back to its place in St. Peter's with all
possible pomp and ceremony. We went in an open
carriage to the piazza, and at the cost of remaining
some hours in the sun, had an excellent view of the
whole procession, which comprised, I should think
pretty nearly the whole ecclesiastical body of Rome
and all the military. But by far the most interesting
object in it was the Pope himself, who walked on
foot immediately behind the relic,—and we saw him
very well; he has a fine and very pleasing countenance, very like his portraits, but he looked pale
and care-worn, with an anxious smile, and indeed
his situation is a most difficult and trying one. I
trust he may be able to hold his ground, but no man
had ever more difficult cards to play. Now as to
our own proceedings.—We left Florence on the 22nd
March, and arrived here on the 30th, having slept at
Arezzo, Camuscia, Perugia, Foligno, Terni, Narni
and Le Sette Vene. It was a very interesting
journey, and we were fortunate enough to have
delightful weather during the whole of it. We
spent a whole day at Perugia, a place with which
Fanny and Susan were much delighted. I cannot
say it captivated me particularly, though many of
the views from it are certainly very fine, and there
are some interesting old buildings. In the way
between Perugia and Foligno we visited Assisi, and
spent some hours in the great church, a wonderful
place it is certainly, and a mine of early art, which

it would take a long time to explore thoroughly. 1848.
Fanny and Susan, who are enthusiastic about the
painters of the 14th and 15th centuries, were in
ecstacies with it. For my part, there was hardly
anything at Assisi that I admired so much as the
portico in the ancient temple of Minerva, in the
Piazza. I was charmed with the vale of the
Clitumnus, and with the country all the way from
Spoleto to Narni, and indeed beyond it, as far as
the limestone district extends,—but above all with
the valley in which the Fall of Terni is situated,
which is even more beautiful than my recollections
had prepared me to find it.   Fanny, who had never
seen a really great waterfall before, and who rather
expected to be disappointed, was in as great delight
with this as I had hoped she would be.   The
deciduous trees around the cascade were not yet in
leaf, so that the dark green of the Ilex predominated,
but the beautiful tall white-flowered Heath (Erica
arborea) was in profuse blossom, and the Judas-tree
putting forth its buds.   I was rather surprised to
see the Maritime Pine growing on the mountains
about Terni and Narni, so far from the sea,—
certainly ths same species that prevails so much
on the coast mountains of Provence and the
Genoese Riviera.   What with the Etruscan walls of
Cortona, the beautiful Thrasimene lake, the Etruscan
antiquities of Perugia, and the pictures of
Raphael and Perugino, the marvellous church of
Assisi, and its beautiful Roman temple, the
Clitumnus, the Velino, the bridge of Narni, the
picturesque towns, the beauty of the country almost

1848. all the way from Arezzo, and the historical recollections connected with every part of it,—this road
is surely one of the most interesting in the world.
But of all the towns and cities of Italy, my delight
is Florence. I was not half satisfied with the sight
we had of it, and would willingly have remained
there till the summer, but Fanny and Susan were
very naturally anxious to come on to Rome, and
our letters had been sent thither. We have taken
lodgings in the Corso for a month, but shall probably stay two months, so that I hope to hear
again from you during our stay Our future movements are very uncertain, owing to the state of
turmoil that all the world is in, which makes it
difficult to know what countries are open to travellers. I fear it would be hardly safe to go to
Naples. We sometimes think of spending the hot
months at Siena, more especially as the Scientific
Congress is to meet there in September, but it
remains to be seen in two months time whether
the roads from hence are safe. I hope we may not
be obliged to return home by sea, but that will
at least remain open to us as a last resource, in case
of a general war.

Fanny and Susan send their love to you and
together with mine, to Emily and Cecilia.

Believe me ever,
Your very affectionate Son,
C. J. F. BUNBURY.

Rome,
April 10th, 1848.

My Dear Edward,

I thank you much for your very agreeable letter of the 9th of March, which I received soon after our arrival here. I was much pleased to read your accounts of the proceedings in Parliament and of the state of opinion in England. The Revolution in France was indeed the most extraordinary event that has happened in our time—far more surprising and less to be expected I think than that of 1830; and though it has been so far unstained with cruelty or wanton bloodshed, yet the course it is taking is such as necessarily makes one very anxious and apprehensive ; its real consequences are not yet half developed, either as respects France itself, or the rest of Europe. But when one sees the administration of such a country suddenly placed in the hands of a set of men, not one of whom is known as a statesman, or has had the slightest previous experience of government—men, some of whom have been notorious as writers for flattering the most dangerous passions of the populace,—when one sees these men beginning their career with making the rashest promises and entering on the wildest and most hazardous experiments of political and social economy,—when one sees the National Guards and the middle classes completely bullied and overridden by the armed populace, and all power falling into the hands of the classes without property and without education—one cannot help fearing very much for the result. I am very glad that there is

1848. such a general feeling in England of the necessity of
letting the French alone, and leaving them to work
out their social experiments for themselves, and it
would seem that, with the exception perhaps of Russia, all the other European powers have too much on
their hands at home to be willing or even able to
interfere with France, but whether France herself
will exercise a similar forbearance in the present
crisis of Italian affairs, appears to me exceedingly
doubtful. Of all the revolutions which springing
out of that of France have succeeded one
another with such bewildering rapidity in almost all
parts of Europe, the one which surprised me most
was that of Vienna, which I had imagined would be
the very last city in Europe to be revolutionized. I
should almost as soon have expected to hear of such
an event in another " Kaiserstadt "—" Pekin."
I must say I rejoice most heartily, in the downfall
of Metternich, and feel a savage satisfaction that
he has not been allowed to escape out of this world
before seeing the complete overthrow of all his
system of policy.

I quite agree with you in looking upon the
virtual abolition of the Treaty of Vienna as a great
blessing to Europe. You will have heard before this
of the outbreak in Lombardy, and of its hitherto
successful progress. We were at Florence when the
news arrived there. I have written to my father an
account of the stir it excited, and as you will
probably see the letter I will not repeat it. All Italy
is now fairly committed to the struggle against
Austria; and the Austrians, notwithstanding their own

Revolution, seems as I expected not at all inclined
to give up their hold so easily. It is reported that
thirty regiments are marching towards Italy. The
Italians, I understand, and especially the Romans,
have a general belief that the Austrians are stupid
and cowardly, and will be easily beaten ; I suspect
they will find it is no such easy job, and that there
is a wide difference between driving the enemy out of
the city and making head against them through a
whole campaign. I most heartily wish for the
ultimate success of the Italians, but I think it will do
them good to have to maintain a serious struggle,
and even to meet with one or two severe checks.
But which ever way we look, the prospect of a
general European war seems more and more imminent, indeed inevitable, and this last new outbreak
that we have heard of,—the rejection of the
authority of Denmark, by Schleswig and Holstein,
introduces a new element of combustion. But what of
Ireland ? Will the physical force party there confine
themselves to rant and frothy nonsense ? Is not the
state prosecution against the leaders of young Ireland
a measure of doubtful prudence ? We are settled
here for the present in good lodgings, in the Corso,
near S. Carlo, a dreadfully noisy situation, but
otherwise convenient. Rome is much less full of
English than usual, a great number having left it in
alarm. The Murchisons are here, but are on the
eve of their departure ; Sir Roderick, of course, full
of horror at the state of the political world. He has
however been very good-natured to us, and much
more agreeable than I ever found him before.

A A

1848. There is plenty of stir here, and abundance of processions and fine shows for the amusement of the
Romans, who are as fond of them as children. On
the 5th there was a particularly grand procession,
when the head of St. Andrew (which had been found
again after having been stolen) was reconveyed to
its place in St. Peters : going in an open carriage to
the Piazza, we had an excellent view of the whole
procession, which was immense,—all the religious
orders, the regular troops, the Pope's Swiss guards,
the whole of the Civic guard, all the Cardinals,—and
what was more worth seeing than all the rest of
the show together, the Pope himself. He has a
very fine countenance, sensible and benevolent, very
like the busts of him which are sold everywhere,
but he looked pale and care-worn. Yesterday again
there was another great procession for two pieces of
cannon, which had been sent as presents to the
people of Rome (I know not from whom) were
brought all through the Corso, escorted by the whole
of the Civic Guard.—You cannot imagine how
rapidly all sorts of reports and stories follow one
another, or how extravagant they often are. One of
the last I have heard is that the Emperor of Russia
has sent an army of 150,000 men to march direct
upon Rome !

Political events now crowd so fast upon one
another, and necessarily occupy so much of one's
thoughts, that it is difficult to avoid filling one's
letters with them, and as I told Fanny, though this
is an interesting time for seeing the Italians, it is a
very bad time for seeing Italy, nevertheless we made

a very interesting and agreeable journey from 1848.
Florence hither by way of Perugia and Assisi,
Fanny's love of mediæval architecture and of the
early painters found abundant gratification in those
towns, and in Arezzo and Cortona and we were all
delighted with the beauty of the country, the
picturesqueness of the towns, and particularly with
the fall of the Velino and the bridge of Narni.
Florence charmed us all in spite of the almost
constant bad weather we had during our fortnight's
stay there; for my part I prefer it to any other town
I have ever seen, and would gladly have spent
months there, as I yet hope to do sometime or
other.

Our future plans are rendered very uncertain by
the disturbed state of the continent, and it is not
impossible after all that we may be obliged to return
home by sea, though it would not be pleasant. We
wish if we can, to spend the hot months at
Castellamare or Sorrento, but this must depend
upon the state of affairs at Naples, for as it is quite
clear that the king of that country is not in the
least to be trusted, one may any day expect a
new Revolution there,—and it may be a bloody
one.

As a specimen of that king's character, it is known
that, while he proclaimed the general amnesty for
political offences, he sent secret orders to his
ambassador at Paris, and to the consuls at the
French ports to grant no passports to any of the
Neapolitan exiles. It is reported that Lord Napier
has advised all the British subjects to quit Naples,

1848. but I do not know what truth there may be in this.
If Naples be unsafe, and yet matters not so bad as
to oblige us to leave Italy suddenly, we may probably
go to Sienna.   You do not mention anything in
your letter about our Uncle Fox's money.   I should
be rather glad to know whether any more of it
has been divided.   I fear that the contents of his
boxes at Genoa sufficiently unsaleable at all times,
will be still more so in the midst of all this turmoil;
but I have heard nothing about them from M. De
la Rue.

We have splendid weather here, and having got
over the first botheration of establishing ourselves in
lodgings, go about sight-seeing pretty diligently;
and Fanny, whose first impression of Rome was
rather unfavourable, is now in great delight.   I
think the Early Christian Basilicas are her especial
favourites, but she also enjoys the antiquities
thoroughly, and studies them with great zeal and
intelligence.

Yesterday (I am now writing on the 12th), we
made a delightful expedition to the tomb of Cecilia
Metella and the grotto of Egeria; and she was
as much struck as I hoped she would be with
the beauty of the Campagna.

Susan too is charmed with the scenery, and
especially with the colouring of the mountains.
I am reading Dr. Schmitz's History of Rome, and
find it a very clear, satisfactory, and useful com-
pendium.   What an improvement upon old Goldsmith's, which we used to read in our schoolroom
days!

Letters will find us here, I hope, till the 2nd 1848. or 3rd of June, and I shall always be glad to hear from you when you have time to write.

Ever your very affectionate Brother,

C. J. F. B.

———

Rome,
April 14th, 1848.

My Dear Mr. Horner,

Susan has, I know, kept you so constantly informed of all our movements and adventures, and of all that we have seen, heard and done, that I shall only try to fill up the picture by touching upon a few points, which she is not so likely to have noticed. I have not, indeed, been able to do much in the geological way, for the constant succession of exciting political events, almost ever since we entered Italy has very much distracted one's attention from quieter pursuits; and moreover, in travelling in this country, especially with two ladies to whom it is entirely new, one's time and attention are necessarily claimed, in the first place by the wonderful works of ancient and modern art, in which it is so rich; thus, I have had but little time to bestow on geology, and I am sure that it is not a science which can be profitably studied by snatches. It is a *tyrannical* study, which requires the devotion of a man's whole time and thoughts, to do anything great in it.

We have found the Murchisons here. Sir Roderick has shown us a great deal of attention and

1848. good-nature. As you may suppose, he is in great
horror at the present political state of Europe. He
seems to have been paying much attention to the
geology of Italy, but he is of opinion that it is very
monotonous, except in the volcanic districts.
Certainly two things are very striking, in what I
have yet seen of Italy—the great uniformity of the
mineral character of the stratified rocks over very
extensive areas, and the extreme scarcity of fossils
except in the upper tertiary beds. Two formations
predominate to an amazing extent :—one, the hard
compact limestone of the Apennines, which sometimes becomes saccharine, and which in Collegno's
map is referred to the Jurassic series ; the other,
what the Italians call Macigno formation, which is
composed of micaceous and often schistose sandstones, of shales, and soft marly-looking limestones,
like those used in lithography. This is the
formation in which Fucoids and Nummulites are
found. Nearly all the Italian geologists regard it as
cretaceous. Savi speaks of this as a point generally
agreed ; but Dr. Perez, at Nice, maintains that it is
Eocene, and says that he has found in it, on that
coast, several species of shells, all of them unquestionably Eocene, except one. In Tuscany, it
would seem by Savi's account that no fossils have
been found in it except Fucoids, Nummulites and
Hamites. Murchison seems to incline to the opinion
that it is Eocene. I have not been able, since I left
Nice, to make out anything bearing on the question
for though we travelled for several days through a
country of this formation, fossils are so rare in it

that the chances are infinite against finding any 1848. in such hasty researches as one can make on a journey. There is no development of the Macigno within reach of Pisa; and during the short time we stayed at Florence there was so much in the city itself to engross our attention, that I had not a moment to spare for geology. I hope, however (if we are not obliged by political events to quit Italy suddenly) to spend some considerable time at Florence, before the winter, and then to have a little leisure for examining the country. I had hoped to find something illustrating this subject in the museums of Pisa or Florence, but I was disappointed. In the Pisa museum, indeed, there is an extensive collection of the geology of Tuscany, but unfortunately, from the manner in which they are arranged, a great part of the specimens are almost invisible, and I saw no fossils from the Macigno formation, except Fucoids, which one cannot rely upon as indicating its age; for the strata in the south of France, which are characterized by the same species of Fucoids, appear to be equally disputable in date. The Florence museum, which is splendidly rich in many departments, appeared to me very imperfect in that of geology—rock specimens excepted; there are, indeed, some local geological collections from the basin of Paris, basin of Vienna, &c.—but what I thought strange—none of the geology of Tuscany itself. It is, however on the whole, a noble collection, and in one department— that of anatomical models—I believe, unrivalled by any in the world; the beauty of execution of those

1848. models is truly wonderful, and I only regretted that I had not science enough to appreciate them thoroughly.

The collection of minerals appeared to me very satisfactory, and very well arranged, but the manner in which it is placed in upright cases along the sides of a gallery which is lighted from one side only is not so advantageous in point of light as that adopted in the British Museum. The specimens from Elba, as might be expected, are very numerous and fine ; the Tourmalines in particular most splendid. One department, which to me was very instructive, was a collection of rock specimens extremely well selected and arranged, and named according to the nomenclature of Alex. Brongniart ; I had been so often puzzled by his terms, that I was particularly glad to find them thus illustrated. There are several interesting things also in the botanical department of that museum, especially wax models of the anatomy of Chara, and others illustrating the process of fecundation in the Gourd, according to the observation of Amici. I suppose you know Collegno's geological map of Italy : as far as I have yet seen, he has marked very accurately the boundaries of the main groups of rocks. Whether there is much yet to be learned in the volcanic districts I do not know ; I rather suppose they have been thoroughly explored ; but I should be very glad to have hints and suggestions as to the points most deserving of attention. I have not yet visited Monte Mario nor the quarries of Capo di Bove, but hope to do so soon, On our way hither, I collected

specimens for you as well as myself, from the fine 1848.
bed of leucitic lava which extends over the country
from Borghetto nearly to Civita Castellana.

What struck me as curious was that *under* this
bed of lava in the ascent from Borghetto and the
valley of the Tiber, there appears a tufaceous conglomerate of scoriæ and ashes, containing large
blocks of the *same* lava in a much decomposed state.
One thing to which I have paid a good deal of
attention in the course of our journey, is the
distribution of wild plants in relation to the nature
of the soil and of the rocks, and I have got together
several observations on this subject, but I
have no room for them here. I am much in
the dark as to scientific proceedings in England,
the "Athenæums" which were sent never reached
us, and that paper is not taken in at the reading-
room here, so that I have seen no account either
of De la Beche's anniversary address, or of any of
the papers read before the Geological Society.
I shall be very glad if you will tell me something
about them.

I have hardly space enough left to touch upon
politics, and indeed I have been writing so much on
that subject to *my side* of the house, that I am
almost tired of it. This is indeed a year of stirring
and wonderful events, and though I do not, like Sir
Roderick Murchison, call the times dreadful, yet
certainly one cannot look without awe and anxiety
at the general wreck of thrones and old institutions
and established powers, which is going on all around
us. I rejoice in the downfall of the selfish Louis

1848. Philippe and of the corrupt and deceitful ministry of Guizot, but the wild career of the government (if government it can be called) which has succeeded them, inspires infinitely more fear than hope. I am very glad that Katharine and her husband did not go to Paris, which I am sure cannot be a pleasant or even safe place for English people at present. I hope Mrs. Power and Mrs. Byrne will not lose everything through the Revolution.

I cannot say I feel quite secure or comfortable even about England, for the financial and commercial difficulties, which cannot fail to be increased by this universal turmoil on the Continent, will give a stimulus to the Chartists, and in Ireland, I fear, terrible events are at hand. Italy has now I trust got rid for ever of the Austrians, but her further progress will for some time to come be a matter of anxiety, for the Italians will find out before long that besides national independence and free forms of government, something more is required to make nations really free and happy.

In addition to the Murchisons we have very agreeable friends here in Mr. Moore and his Son (brother and nephew of Mr. Carrick Moore), but we have so much to see that we none of us wish for *much* society.

My love to Mrs. Horner and Leonora and Joanna and to both pairs of Lyells.

Ever your very affectionate Son-in-law,

C. J. F. B.

Rome,
April 30th, 1848.   1848.

My Dear Mary,

We have now been here a month, and I do
not think either Fanny or Susan have been disappointed in their expectations of Rome ; we have been
tolerably active and busy in sight-seeing, but I have
a great deal more to see, and I dare say we shall
not find the next month at all too long.

I am very sorry that Susan has determined to
leave us. I can quite understand her feeling of
home-sickness, and her wish to see a good deal of
Katharine before she goes to India. We were
afraid Susan would be leaving us in a week's time,
but it is now much more comfortably arranged, as
she is to stay till the 1st of June, and then is going
in a very good steamer, and in company with some
friends who have been very kind to her here. I
hope her Italian tour will not have been quite
unproductive or unprofitable to her.

We have had some very pleasant acquaintances
here, but in a little while there will be none left
except Mr. Gibson, for the Murchisons are gone,
Mr. and Mrs. Hutchings are going to-morrow, and
Colonel Moore in a few days, and in short, all the
English, as usual, are taking their flight in haste, as
if it were dangerous to ren..ain at Rome after Easter.
The number of English here, this season, has
indeed been much less than usual ; and the shop-
keepers, especially those who deal in ornaments and
works of art, feel seriously the falling off in their
custom ; but if this general turmoil on the Continent

1848. continues, or (as seems likely) increases, they will
suffer much more next winter.

Rome, itself, has been quite quiet during the
the last time of our stay,—free, I mean, from any
disturbance, though agitated by a thousand rumours,
for the Romans are the greatest newsmongers and
swallowers of marvels that can be, and every day
brings forth some new story, generally fabulous, and
often absurd, but always eagerly received, and
believed for the time.   One cannot indeed wonder
that there should be great excitement among a
people of so ardent a character as the Romans,
when the events which are really taking place in all
parts of Europe are of so exciting a character, and
especially when their fellow-countrymen are engaged
in such a struggle with the Austrians.   Rome is now
governed in reality by the Clubs and the Civic
Guard, the Pope having no power but to follow the
impulse which they give.   The Civic Guards, however, who are quite *the rage* here, as well as at
Florence) are really very useful, and perform the
duty of a police quite as efficiently as it was ever
performed under the old régime.   We are living in
the Corso, in the heart of modern Rome, in a
situation convenient in many respects, but horribly
noisy; it is indeed the noisiest city that I know
except Rio de Janeiro.   Modern Rome indeed, is
be no means an agreeable city, nor should I call it a
beautiful one, except in the general view from any of
the surrounding heights; but the ancient and
deserted parts of Rome are delightful, and so are
the villas around it.

I should never be tired of rambling among the 1848. noble masses of ruin, which cover the Palatine Hill or among the gigantic ruins of Caracallas Baths or the green lawns and beautiful pine groves of the Borghese and Pamfili villas; nor of the delicious view of the Campagna and the mountains from before S. Giovanni Laterano. Nothing can be more interesting than the lonely Campagna,—a green desert, varied by fine undulations of the ground, abrupt ravines and miniature cliffs, spangled with wild flowers, undisturbed by any modern vestige of the operations of man, but dotted over with ancient Roman tombs, some entire and some in ruins, and traversed by long lines of grand arches, the remains of the magnificent aqueducts which formerly brought to Rome so vast a supply of water. And then the mountain boundary of the Campagna is so fine, especially the volcanic group of the Alban hills, which is about the most beautiful mountain group that I know ; and the glittering white towns on their sides have such a gay and bright effect in the landscape, contrasting with and relieving the sublime solitude of the Campagna.

The spring is now in all its beauty here, the trees in full leaf, and the public walk perfumed by the acacia blossoms. The ruins, especially those on the Palatine are covered with most beautiful wild strawberries and thickets of Mastic, Terebinth, Laurustinus, Phillyrea, Coronilla and a variety of other handsome shrubs, intermixed with a gigantic plant of the fennel kind (ferula commensis), which at a little distance looks like the flowering stem of

1848. the American Aloe.   In the grass lawns of the villas,
and on the green slopes of the Campagna, grow a
profusion of Orchises, either strangers to England or
very rare there, such as Orchis tephrosanthos and
Ophrys aranifera,— a variety of cruciferous plants
and grasses, and many other interesting herbaceous
plants.   The aspect of the vegetation is perhaps not
on the whole so exotic as in Provence and about
Nice, but when one comes to examine it, the
difference from the vegetation of our country is
striking, so many of the common plants of Rome,
and of those which play an important part in the
vegetable covering of the surface, are unknown
among us.

In my list of 42 species found here during this
month (not including any of those which are very
common everywhere) I find that 25 are completely
strangers to England, and 7 or 8 others, though
included in our Flora, are very rare with us.   Some
of the most common grasses of Rome are strangers
to England, such as Bromus Ligusticus, Aloepe-
curus utriculatus, Secalæ villosum.   There is a
very good descriptive Flora of this region, by
Sebastieni and Mauri, (both of whom are since
dead) it contains 1200 species of flowering plants,
and the descriptions are as good as in any work
of the kind that I know ; no doubt the list might be
much enlarged by anyone who had leisure and
opportunity to explore the recesses of the Roman
Apennines and the marshes of the coast ; unfortunately much of the territory is dangerous to explore
during the summer.   I have done nothing here in

the way of geology, except that I have learnt a good 1848.
deal from conversations with Murchison, and that
he took me to see the collections of Professor
Pouzi, where I saw several interesting things—
vegetable remains in fine preservation, and seemingly of recent species, in a freshwater limestone or
travertin (not of the present era) from the banks of
the Trouto near Ascoli; and plants apparently
identical with those of Aix in Provence (in particular
leaves of a Cinnamomum) found in company with
fossil fish in the territory of Sinigaglia. But the
most remarkable things were Dicotyledonous leaves,
exceedingly like those of the common Chesnut,
found in the Macigno sandstone (which has been
generally considered Cretaceous) on the Adriatic
side of the Apennines, in Picenum.

These troubled times are exceedingly in the way
of one's exploring the country with any scientific
view, and they render it impossible or useless to form
anything like positive plans for one's movements for
any time in advance. We have however pretty well
made up our minds (barring accidents) to spend the
hot months at Sorrento, of which we hear most
delightful accounts from everybody.

I hope Lyell will write to me a particular and
precise statement of the questions which he wants
solved at the Temple of Serapis, or anywhere else near
Naples. I am glad to hear he is going on so well with
his book; I suppose it will come out next winter.

I hope you will write to us often after Susan
leaves us, for we never hear from my side of the
house. Our plans, after Sorrento are quite uncer-

1848. tain : I often feel an unwillingness to look forward to
a second winter abroad, and I think Fanny too
sometimes feels a certain degree of hankering for
home, but whether we really shall return home this
year, will depend upon many contingencies. It is
uncertain too whether we shall be able to go to
Turin ; the last I heard from that quarter was that
the roads between Genoa and Turin were so
broken up by artillery, &c., as to be almost impassable for carriages.

May 3rd.—We have been on the very brink of a
Revolution here, but I have no room to tell you the
particulars, and no doubt Susan has written them
all home. My love to your husband and to
Katharine and her husband and the Rivermedians.

> Ever your very affectionate,
> C. J. F. BUNBURY.

———

> Rome,
> May 4th, 1848.

My Dear Father,

I sit down to write to you on your birthday. I trust you are enjoying it in good health, and
free from anxieties about the health of any who
are dear to you,—that you may live to see many
more birthdays, and may very long enjoy health
and happiness, and the honour, love, obedience,
which are so justly due to you, is my most
earnest and heart-felt wish and hope.

I have allowed a month to pass without writing
to you, because I had nothing of any consequence

to tell, but now I have matter enough, for within the 1848
last few days we have been on the very brink of a
Revolution here, and one which might have been felt
throughout Catholic Europe.   The storm has blown
over however, and all is quiet again.   But I must go
back some little way to explain the matter.   I mentioned in my last letter that at the same time that the
King of Sardinia and the Grand Duke of Tuscany
put their forces in motion against the Austrians,
the Pope also sent part of his regular army, and a
great number of volunteers to the frontiers, and every-
one supposed that he intended to adopt the same
decided course as the other Italian sovereigns.
But it seems that when he addressed the volunteers
and gave them his blessing on their departure from
Rome, he enjoined them to go only as far as the
frontier, and *not* to cross it.   The volunteers however
did not believe him to be in earnest (nor, I fancy,
did anyone else at the time) ; they crossed the
frontier in great numbers and joined the other
Italian armies which are now actively engaged in
war against the Austrians.   But in the meantime
the Austrian Ambassadors remained at Rome, and
the two powers continued ostensibly at peace with
each other, and accordingly the Austrian Generals
made it known that they considered the Roman
volunteers as brigands, who were waging a war
unauthorized by their own Government — that
they would treat them accordingly, and hang without mercy all that they could catch.   It is said that
a young painter of the name of Caffi, who was one
of the volunteers, having been wounded and taken

1848. prisoner, was actually found a short time after, hung
to a tree. This may be one of the fabulous reports
which circulate here in such amazing abundance,
but however this may be, the avowed intention of
the Austrians and the knowledge of the fate to
which the volunteers were exposed, caused great
excitement here, and the people loudly demanded
that the Austrian Ambassador should be sent away
and war declared in form.

On Saturday, the 29th, the Pope made a long and
rather weak and washy address to the Consistory,
the pith of which was, that it was contrary to his
conscience, and to the laws of the Church, to wage
war against anyone, except in self defence, that he
had only allowed the volunteers to go to defend the
frontier, and could not sanction any offensive movement on their part.

(May 5th). Since I began this, I have had the
very great pleasure of receiving your kind and
interesting letter of the 22nd, but before I proceed
to answer it, I will finish my story about the events
of the last week :—

When the Pope's harangue to the Consistory, was
published (late on Saturday), it excited an immense
sensation ; the people, especially those who had
relations or friends among the volunteers, were
exceedingly irritated ; the Clubs *(Circoli)*, which
have more power here than any of the regularly
constituted authorities, held sittings all through
Sunday, debating on the measures to be taken.
The Ministry entreated the Pope to allow them

to declare war, offering their resignation as an 1848.
alternative; he accepted their resignation.

A deputation of Prince Corsini, Prince Doria, and
Count Mamiani, went to the Pope, to impress upon
him the intensity of the popular feelings on the
subject of the war, and the necessity of yielding to
it; he would yield nothing. The Civic Guard took
possession of the Castle of St. Angelo, the powder
magazines and the gates of the city. All through
Sunday the Corso was thronged with people (chiefly
of the upper and middle classes) gathered together
in groups and knots, discussing the state of affairs
with all the vehemence of Italians.

On Monday, the Pope still continuing firm in his
resolution against war, the agitation went on increasing, and as usual at such times, it was inflamed by
wild rumours of plots and treasonable correspondence. One Cardinal was stopped, it is said,
while attempting to escape from Rome in disguise
(though some say it was his brother who was taken
for him) and one or two others were placed under
arrest in their own houses by the Civic Guard. The
gates were strictly guarded and no one was allowed
to go out, even for a walk or drive in the country,
without an order from the head of the police.

Towards the evening we were told that the Pope
had abdicated his temporal power, but this was not
true. Still there was no riot or actual disturbance,
but the people began to use very violent and
threatening language against the Cardinals, and
talked openly of cutting their throats. Early on
Tuesday morning the Pope issued a proclamation,

1848. expressing his horror at the sanguinary and ferocious
language which had been used against those
venerable personages, asking whether this was the
return for what he had done for the good of his
people, and threatening to use his spiritual power—
in fact to excommunicate the Romans. This had
some effect, for though the chiefs and the members
of the Clubs might, themselves, laugh at excom-
munication, they knew that it would tell with
formidable weight upon the populace.

Accordingly all remained quiet on Tuesday ; and
that evening it became known that Count Mamiani
had been appointed Secretary of State, and com-
missioned to form a new Ministry. This gave great
satisfaction. Mamiani who was for many years an
exile, and indeed, has but very lately returned to
Rome, is a very learned and able man, and generally
admitted, even by those who are not favourable to
the Liberals, to be a very honest one. We saw him
at Genoa, and thought very well of him, as far as it
was possible to judge in a single conversation.

The last two days have passed in negociations,
and the new Ministry is not yet definitely formed,
but the paper of yesterday evening, says that the
principle difficulties have been overcome, and that
the composition of the Ministry is nearly settled.
For almost a week now, Rome has in effect been
governed by the Civic Guards, and I must say they
have acquitted themselves admirably, preserving
order, and peace and discipline, so that with all the
agitation, there has been nothing like riot, and
nothing to alarm or annoy strangers. But I

certainly expected to have to write to you that the 1848.
Pope had resigned his temporal authority, and was
reduced to a mere Bishop of Rome ; and I am not at
all sure that it may not yet come to that in the end.
I am very sorry for Pius, who is certainly an excellent
and most benevolent man, far more enlightened
than the generality of Popes, and who set the
example of reform to his brother sovereigns in Italy ;
and his position has been a most difficult and
embarrassing one ; but I cannot help thinking that
his conduct in this volunteer affair has been very
weak and short-sighted, if not very insincere.

One thing struck me much in this agitation and
quasi-revolution—that it seemed to be in no degree
the work of what we call *the mob*. The assemblages
that crowded the Corso were composed almost
entirely of well-dressed people, of what are called in
England, the *respectable* classes, and the great nobles
of Rome — Doria, Borghese, Corsini, Rospigliosi,
take a leading part in all their movements. The
mechanics and the peasants do not seem to interfere,
but if we may believe what is said by some of the
English, they are rather attached to the priesthood,
and to the old order of things. The man here who
has most power over the lower classes is a certain
Angelo Brunetti, commonly known by the name of
Ciceroacchio, who by the accounts we have heard of
him, is a remarkable man. He was of low origin,
but became wealthy through his own industry and
exertions, and has acquired a vast influence over the
common people, not by his talents or eloquence, for
I am assured that he has no conspicuous ability,

1848. but by his reputation for goodness, generosity and integrity. He seems to be a man uncommonly disinterested and free from personal ambition, has refused various offers of place and profit, and shows none of the usual vices of a demagogue. The Clubs avail themselves of his influence to recommend to the lower classes any measures which they wish to carry ; but he is said to be strongly attached to the Pope, and always anxious to know whether *he* approves of the measures which are recommended by the Clubs. Brunetti wished to go with the volunteers, but he is of so much importance here, that the Government entreated him to stay to keep the people in order ; so he sent his two sons.

(*May 6th*). The new Ministry, with Mamiani at its head, is at length definitively formed, and of course, war will speedily be declared against Austria—Prince Doria-Pamfili is Minister of War.

I was much interested by your remarks on the operations of the Sardinian King.

The war seems to be going on rather slackly at present, but it appears there are frequent skirmishes, in which, according to the Italian accounts *they* are always victorious ; but they are so grandiloquent, and make so much of every trifling affair that it is difficult to get at the true state of the case. It is known however, that Udine has capitulated to your friend General Nugent.

It is said that a short time ago, three couriers arrived in succession with despatches for the Pope from the seat of war, but their contents were kept

strictly secret, and this added much to the anxiety 1848.
and irritation of peoples' minds.   Last night there
was a report that Charles Albert had gained a
victory, but there seems to be no certainty of it.
I was delighted to read the accounts of the failure
of the Chartists in England ; it was indeed a fine
display of the spirit and resolution of the most
important part of the English people, and of their
attachment to law and order ; and must have had
a great effect upon the French, who certainly
reckoned upon London following the example of
Paris.   It was in truth a great and important
victory, though gained without a blow, and it was
happily followed up by the discovery of the mass of
sham signatures attached to the famous petition,
and the enormous impostures practised in regard
to it, which must have damaged the Chartists and
their parliamentary advocate very much.   As for
Ireland, I can hardly imagine how matters can end
there without an outbreak : whether O'Brien and
the rest of the villainous demagogues are in earnest
or not, their wretched dupes certainly are, and they
have as you say, gone too far to retract.   I hope
the leaders will not escape their fair share of
punishment.

We were very happy to hear that Cecilia was so
much better, and to have such a favourable account
of little Harry.   Fanny has this morning received a
comical but very satisfactory letter from the monkey
himself, and a long one from his excellent mother,
who writes cheerfully, though she laments that
Hanmer is so incessantly occupied and worried

1848. by business and cares of various kinds, that he never has any leisure for the quiet enjoyment of home.

We were both much shocked to hear of the death of Lady Katherine Jermyn : I feel very much for her poor husband, and her loss will be much felt by all the neighbourhood.

Poor Lady Mary Fitz Roy's death I had seen in the newspapers.

We have taken our lodgings here (36, Via de' Pontefici) for another month, to the 3rd of June. There is a vast deal to interest and occupy us here in Rome, but yet I shall hardly be sorry when the time comes for us to leave it, for the noise is distracting, and both Fanny and I feel a good deal the depressing and enervating influence of the climate.

We have almost determined upon spending the hot months at Sorrento, of which we hear delightful accounts from everybody. Susan, I am sorry to say, is going to leave us ; she is anxious to get home, and has made up her mind to return by sea in a steamer which leaves Civita Vecchia for Southampton, the beginning of June, and in which she will have the companionship of two ladies with whom she has become very intimate here.

We have had some pleasant acquaintances at Rome, particularly Mr. Francis Moore and his son Colonel Moore, Mr. Gibson the sculptor, and Sir Roderick Murchison ; but the latter is now gone, and Colonel Moore also.

Fanny sends her love to you, and unites with me 1848.
in love to Emily and Cecilia.

Believe me your very affectionate Son,

C. J. F. Bunbury.

P.S. — Fanny abuses Cecilia for never writing
to her, and thinks she deserves better treatment.

———

Rome.
May 22nd. 1848.

My Dear Father,

I begin a letter to you, to be sent off when
we are leaving Rome. We are again thrown into
a state of great uncertainty about our movements
for this counter-Revolution at Naples has made it
very doubtful whether that place will be safe for
strangers, and has consequently deranged all our
plans. The popular party at Naples has behaved
most foolishly, put itself completely in the wrong,
and drawn upon itself a defeat which may have an
injurious effect upon the cause of freedom in Italy
generally. Though it is very disagreeable with
the opinion we all have of the king of Naples,
to be obliged to think him in the right, I cannot
help feeling that he was so in this instance. The
Neapolitan constitution, which was granted no
longer ago than last February, was abundantly liberal
in its provisions, and indeed was thought at the time
so liberal, that some said the king's object was to
outdo and spite the Pope! yet before it had even
come into operation, the people had grown dis-

1848. satisfied with it, and wanted something else.—The chambers were to meet last Monday, the 15th; the constitution had expressly reserved to the king the right of nominating the peers, and he had done so, but when the list of peers was published, the popular party disapproved of it, and without waiting for the meeting of the Chamber, sent a deputation to the king to demand that this part of the constitution should be annulled, and that there should be no Chamber of Peers at all. This modest demand was supported by the whole of the Civic Guard, and by the Ministry; yet the king remained firm, temporized till he had made his arrangements, and then gave a flat refusal to the last deputation that was sent to him. This was on Sunday evening, the 14th; the Civic Guards employed all that night in raising barricades, and the fighting did not begin till near noon on Sunday. The king wisely employed only his Swiss Guards and artillery, and moreover gave them orders not to fire until they were fired upon. The Swiss began to remove the barricades! they were fired at from the windows, then they attacked, carried the barricades, stormed the houses, and before night had put down all resistance. The Civic Guards and the liberals in general seem to have shown as little courage in the fight as they did moderation before it. They have brought themselves into a desperate scrape; the king has suppressed the Civic Guard, obliged all the citizens to give up their arms; dismissed his Ministers, put off the meeting of the Chamber, and in short has the game in his own hands ; and now that he has found by experience how 1848. well his Swiss are able to defend him against a popular insurrection, I am afraid he is not likely to use his power with moderation or discretion. The Lazzaroni sided with him, and plundered a great many houses, and committed many acts of violence ; the Roman newspapers as usual, exaggerated these outrages, and represented them as a deliberate sacking of the city after the fighting was over, but it appears that the Lazzaroni followed the Swiss, and entering the houses which the soldiers had stormed in the progress of their attack, took advantage of the opportunity to plunder.

According to a letter which Mr. Macbean (our banker here) received from an English gentlemen at Naples, no houses were plundered, except those from which shots had been fired at the troops. With all this, Naples is not likely to be a very peaceful or agreeable place at present, and we are much afraid that we shall have to give it up.

Our project at present is, to go to Albano on the 1st of June, to stay a week there and at Frascati, and according to the news we receive from Naples, to make up our minds by the end of that time, whether we shall go southwards or to Siena.

Susan, I am sorry to say is going to leave us ; she has become home-sick, and is unwilling to wait for the uncertain time of our return, and so she has determined to brave the disagreeables of a fortnight's voyage, and is going home from Civita Vecchia in the Ariel steamer, in company with some ladies

1848. with whom she has formed a friendship here. I
heartily hope we shall return home before the
winter, but it depends a little on circumstances,
for France and the Rhine *may* be too much disturbed to allow of our returning by land, and I
should be unwilling to expose Fanny to a stormy
voyage late in the autumn.

(*May* 24*th*). It is said that the insurrection at
Naples was produced by emissaries who went from
hence with the express object of stirring up the
popular party to a Revolution. Many such
emissaries it is said are spread through Italy,
employed by the notorious Mazzini, and by the
ultra-republican or anarchical party in France, and
they have been trying to produce Revolutions and
to overturn monarchy at Genoa, Florence, and
other places. Hitherto, fortunately, they have not
been successful anywhere, but their machinations
add very seriously to the difficulties of the great
questions which the statesmen of Italy have to
solve. The rational liberals here, such as Dr.
Pantaleone, deeply lament the folly of the
Neapolitans, who have endangered the good cause
all over Italy. I must own I have much less
sanguine hopes for the cause of Italian freedom,
than I had a few months ago ; it seems to me that
the Italians have much less to fear from the
Austrians than from themselves. Not but that
there are, among the liberals of this country, many
wise and good and well informed and thoughtful
men, but they cannot control the popular impatience ;

a pack of rash, headstrong, thoughtless, impetuous 1848.
young men, like "Young France" and "Young
Ireland" keep the people in a state of perpetual
fever and restlessness, and call it "liberty" to have
every thing their own way, and to silence all
attempts at opposition.

(May 26th). We hear from Dr. Pantaleone that
the king of Naples has recalled his troops from the
war against Austria, bnt that they have refused to
obey, and have placed themselves under the orders
of Carlo Alberto. The Sardinians are bombarding
Peschiera, and the powder-magazine of one of the
forts which constitute the system of defence of that
place, has been blown up, so that the surrender of
Peschiera is expected in a short time.

The Roman parliament is just about meeting,
and Dr. Pantaleone (who is himself a deputy) thinks
that it is well constituted, and will do good.—We
are kept in such a constant state of excitement
about politics that it is difficult to avoid filling one's
letters with what one hears, yet for my own part I
am quite weary of the incessant turmoil, and long for
peace and quiet, so that I shall leave Rome this
time with very little regret. The weather too has
disagreed with both Fanny and me. This has been
a month of sciroccos, and damp, cloudy oppressive
days, varied by frequent thunderstorms and heavy
falls of rain, with little of really fine weather. We
have seen a good deal of Mr. Gibson and Mr.
Penry Williams, and like them both, though Gibson
especially, is a great Tory. His works are really

1848. beautiful, and his conversation, (particularly when one can keep him off the subject of politics, which is difficult), is clever, instructive and interesting, and bears strong marks of an original and vigorous mind. He accompanied us the other day to the Villa Albani, which I never saw before, it is rich in beautiful and interesting works of ancient art, and contains some of first-rate excellence such as the head of Antinous in bas-relief, and the bronze Apollo, which Winckelman believed to be an original of Praxiteles.

We have been pretty assiduous in our visits to the Vatican, where Susan has revelled in Raphaels, while I have given much more of my time and attention to the sculpture-galleries, and Fanny has divided herself pretty equally between them. I certainly feel that I take more delight in sculpture than in painting, perhaps partly because of its connection with classical antiquity, and because the turn of my mind is decidedly much more classical than *mediæval*. But I cannot help thinking that the art of sculpture has been brought to a higher degree of perfection than that of painting, it is true that the works of the great painters of antiquity are lost, but I have scarcely ever seen a picture, even of Raphael, that did not leave a good deal to be desired, whereas I have seen many ancient statues that appear to me in their way absolutely perfect.

(*May* 29*th*). We have heard nothing more from Naples, it may be inferred that things are quiet

there, but we are still left in a state of great 1848.
uncertainty.   All that we have decided upon, is to
go to Frascati, and return to Rome, after about a
week.   Here all is quiet and expected to continue so.

With our love to Emily, Cecilia and Harry,
believe me,

Ever your very affectionate Son,

C. J. F. BUNBURY.

JOURNAL.

June 2nd

Rome to Albano.

June 3rd.

Albano.

The day cool and windy, but clear and very
enjoyable.   Fanny and I spent the morning most
agreeably among the woods about the beautiful lake.
We walked up to the ridge by the Capuchin convent
from which we looked down on the lake ; then
descended by pleasant paths among the copsewood,
admiring the beautiful wild flowers which grew
everywhere among the bushes.   Fanny sat down
and made a sketch of Monte Cavo and Rocca di
Papa, while I strolled about and collected plants.
I found that magnificent plant the Orchis hircina
(Satyrium hircinum of Linnæus), and various
others.   The crater-form of the Lake of Albano
is very striking.   Its rim is unbroken, but is lowest

1848. in the north-western part, between Castel-Gandolfo
and Marino, where also it descends to the lake with
a comparatively easy slope.

About the eastern and south-eastern part of the
circuit, from Palazzuolo to about the Capuchin
convent of Albano on the one side, and to a point
nearly opposite to it on the other side, the hills
forming the border of the crater are much higher,
and descend to the water with a very steep and
uniform slope; in this part too, they are most
densely and uniformly wooded.   The outline of
their crests is not much varied ; but the great
convent of Palazzuolo stretching along the hill-side
amidst the woods, the bold mass of Monte Cavo
(Mons Latialis) rising above it, and the town of
Rocca di Papa picturesquely clustered about a
prominent crag on the left of this mountain, give an
interest to the scenery of that extremity of the Lake,
as the town of Castel Gandolfo, with its conspicuous
cupola, does to the western parts.   The hills are
everywhere green.   There are some ledges of rock
below Palazzuolo, and further to the N. W.
beneath the site of Alba Longa ; but on the whole,
the proportion of bare rock or soil exposed is very
small.   After dinner we took a very pleasant drive,
passing along the beautiful road from Albano to
Castel Gandolfo, and thence to the Capuchin
convent, under the shade of magnificent oaks and
ilexes ; then along a pleasant shady lane between
rocks and trees, from the Cappucini to Ariccia, and
thence by the high road to Gensano, where we
looked down on the Lake of Nemi, now dark and

sombre from the shadows of evening ; and so back 1848.
to Albano by the post road. A profusion of beautiful wild flowers in the woods and on the rocks
by the road-side,—in particular, Silene Armeria,
with its bright rose-coloured blossoms, the large
yellow-flowered milk-wort (Polygala flavescens), and
numerous Vetches.

We looked down on the fertile and cultivated
basin below l'Ariccia,—the bed of a volcanic lake,—
saw Monte Giove, the supposed site of Corioli, and
passed by the curious monument commonly called
the tomb of the Horatii and Curiatii, but now
believed to be that of Aruns, the son of Porsena.

June 4th.

Walked again among the woods above the lake.
Fanny made a sketch of Castel Gandolfo. I
rambled in the direction of Palazzuolo. The Oaks
and Ilexes along the ridge between Castel Gandolfo
and the Capuchin convent above Albano, are superb ; but elsewhere there are few large trees
around the lake ; the wood which clothes the
interior slopes of the crater is chiefly coppice, of
Oak and Chesnut, Hazel and Hornbeam, partially
mixed with Holly, Dogwood, Ilex and Laburnum. One
of the prettiest, as well as of the most abundant, of
the many charming wild flowers which grow amidst
these thickets is the Orobus vernus, with its rich
bunches of flowers of various shades of crimson and
purple, delicately veined. Scutellaria columnæ, a
very stately and handsome labiate plant, which
I never saw elsewhere, is not unfrequent in the
woods. c c

LETTERS.

My Dear Father,

1848.   Our plans have been all overset by a most
unlucky disaster, which has befallen poor Susan, so
I write once more from Rome to give you an
account of it.—

Susan left us on the 1st to go to Civita Vecchia,
to embark on board the "Ariel" with her friends,
Mrs. Heywood and Lady Belcher, with whom
she had settled to return to England. We went the
next day to Albano, where we spent two delightful
days, and intended to spend many more ; when on
Monday we received the intelligence that the
"Ariel" had run on a reef ten miles from Leghorn,
and that Susan and the other passengers were safe
indeed, but prisoners in the Leghorn Lazzaretto,
where they were likely to be detained twelve days.
They must have been in a miserable plight, for,
independent of the terror, they had been several
hours in an open boat in their night-clothes,
drenched to the skin, and, on arriving at Leghorn, were immediately placed within the bare
walls of the Lazzaretto, where they remained
without food or change of clothing, till the English
Consul came to their assistance. Most of the

passengers were ladies; fortunately they were not very numerous, or they might have had much difficulty in making their escape. The accident seems to have been caused entirely by the abominable negligence or stupidity or rashness of the captain, for the vessel struck in broad daylight (at six in the morning) in fine weather on a reef of rocks, which I am told is perfectly well known and laid down in all charts. This disaster has completely knocked all our plans on the head, for we feel it to be our duty to proceed immediately to Leghorn to comfort and assist poor Susan, who, even if her health does not suffer, which we have some fears of, is likely to have a melancholy time of it in the Lazzaretto.

We were quite charmed with the scenery, the air, and the quiet of Albano, even during the short time we were there, it had already begun to do us both decided good. Rome, on the contrary, is detestable at this season, with its stifling heat and its noise and dust, and we are longing to escape from it.

The very day that I had sent off my last letter to you, I received your very kind and interesting letter of the 17th and 18th of May. Barton must have been lovely at that time.

I have no further political news to send you from Italy, except that Peschiera is taken, and that the Italians under Carlo Alberto have repulsed the Austrians with heavy loss it is said, in a severe action fought at Goito; but I have not seen any particulars. This is good news however, and gives hope for the ultimate success of the Italians; as if

1848. they can hold their ground against Austria without foreign assistance, as at present they seem able to do, we may hope that the French will not interfere. The National Guards at Paris seem to have done their duty well on the 15th, and it is a good thing that those vile wretches the anarchical faction should have shown themselves openly and been strongly put down. I hope the Government and the assembly will have courage to punish them with vigour. The state of France is certainly very anxious and perilous, but the National Guards both in Paris and the Departments have shown so good a spirit and seem to be so effective and powerful a body, and the elections too have shown the moderate and national party to be so strong, that one may hope all will come right in the end, though after many struggles.

I am very glad to hear that Mr. Rickards liked my book. Pray remember me kindly to him and to his family the next time you see them.

It is amazing what a variety of wild flowers, especially of the Vetch tribe, we found in those beautiful woods of Albano ; novelties actually crowded in upon me faster than I could examine or preserve them. I do not remember ever before to have seen the Laburnum growing wild as it does there. I wonder we have not more of the beautiful flowers of Italy in common cultivation in our gardens ; they seem to have been all thrust aside by the Californian novelties, which are not at all superior in beauty.

———

June 9. We have at last heard from Susan. She 1848. was released from quarantine on the 5th, and dear, good William Napier, who was at Florence, went down to Leghorn as soon as he heard of the accident, and persuaded her to travel with him and his wife to England. But by a new piece of ill-luck, poor little Cissy fell ill just then, and obliged them to put off their departure, so Susan, who is most impatient to get home, returned to Leghorn to proceed to England with Lady Belcher.

I will write again in a few days.

Believe me your very affectionate Son,

C. J. F. BUNBURY.

————

Siena,
June 15th. 1848.

My Dear Father,

I wrote to you just before we left Rome, giving you an account of poor dear Susan's disastrous shipwreck. The day after I had sent off my letter to you, we received the missing letters from her, giving all the details of the accident, and a truly frightful one it was. We left Rome on the 11th, as there was a chance that Fanny's letters might have induced Susan to wait for us, and we knew we should hear from her at this place. Consequently we made the journey in a very hurried manner, sleeping only at Viterbo and Radicofani. We travelled post which we shall certainly not do again in Italy if we can help it, for vetturino travelling is preferable in every respect, unless one is in a great hurry.

1848.   We mean to go to Florence the day after to-morrow.

Although not so full of interest as the Perugian Road, that from Rome hither well deserves to be travelled in a more leisurely manner than we could do this time. We were charmed with the beauty of the Lakes of Vico and Bolseno, and indeed of the forest country all the way from Ronciglione to the Papal frontier.

We rested for some hours at Bolsena. I went to examine the fine basaltic columns, which interested me much. I regretted not being able to spend a day at Radicofani, for, wild and desolate as it is, the mountain would be interesting to explore, in respect both to its geology and botany. The clay country between Radicofani and this place reminded me much of some parts of the south of Sicily; it is much abused by travellers in general, and I dare say that in winter it is exceedingly dreary and dismal, but at this season, from the singularity of its aspect, and the variety and beauty of the wild flowers, I thought it interesting. Indeed, throughout the journey we were delighted with the wild plants—the Spanish Broom, the Everlasting Pea, the purple and white Cistuses, the profusion of white Roses, climbing over the hedges and thickets and the Venus's Looking-glass in the corn-fields.

Siena pleases us much, and I am particularly struck with the pictures of Razzi, who seems to me one of the most admirable of painters, though in general so little known or talked of. Before coming hither I knew him only by his frescoes in the

Farnesina palace at Rome, which always charmed 1848.
me, and by one small picture in the Chevalier
Kestner's collection; but from his " Resurrection,"
which is now in the Sala del Gonfaloniere in the
Palazzo Publico here ;—his " Holy Family," in the
chapel; his " Three Saints," and "*putti*," in the
Sala del Consiglio, and his grand figure of Christ,
bound to the column in the Academy,—I should
rank him as inferior to none but Raphael. Yet, out
of Siena, one rarely hears of him.

Our present plan is to remain at Florence till
about the 14th July, then to proceed by Genoa to
Turin, where I wish particularly to examine the
fossil plants from Tarantaise in the collection of
Professor Sismonda.

Fanny sends her love.

Believe me ever your affectionate Son,

C. J. F. BUNBURY.

Florence,
June 20th, 1848.

My Dear Mr. Horner,

I think I need not tell you how much
Fanny and I have been grieved by the disaster
which has befallen dear Susan since she parted
from us.

We left Rome for this place on the 11th, and
came by way of Siena, a road which was quite new
to me. The country between Rome and Siena is
divisible in a general way into three natural regions,
well marked by their physical features:— 1.—The
Campagna of Rome. 2.—The region of volcanic

1848. hills and lakes and woodland from Ronciglione to
Acquapendente   or   indeed   to   the   frontier   of
Tuscany; and 3.—The   country   of   tertiary   clay
hills, from Radicofani to Siena.   Of course these
regions are not abruptly limited, but in some degree
shaded off one into the other; the transition especially
from the first into the second, both being of a
volcanic nature, is rather gradual.   This northern
part of the Campagna, is much more hilly than
those portions on the other sides of Rome, and has
a good deal of wood here and there, especially in the
numerous ravines, yet still it may be called in a
general sense, an open undulating plain, at least as
compared with the country which succeeds it on the
north.   The second region is wild and very beautiful
—a tract of irregular volcanic hills, richly clothed
with woods of ancient growth, and enclosing lovely
blue lakes, lying in crater-like basins.   Here were
the Ciminian forests, looked upon with such awe
and dread by the Romans, in the early times of the
Republic ; The Ciminian Lake, now called Lago di
Vico, a charming little sheet of water, deeply em-
bosomed in the hills ; the larger lake of Bolsena,
still more beautiful, though of a less peculiar
character ; the little town of Bolsena, on the site of
the ancient Volsinii, the especial seat and centre of
the religion of Etruria ; and the picturesque towns
of Montefiascone and Acquapendente.   All the
scenery of the Lake of Bolsena is lovely.   The first
views as one looks down on it from the hill of
Montefiascone ; the whole descent from thence to
the shore, and along it to the opposite extremity of

the lake; the ascent to S. Lorenzo, and the view 1848.
looking back from thence. A little way before one
reaches Bolsena, there is a fine display of prismatic
basalt in the rocks overlooking the road; I walked
back to examine it, and procure specimens, while
Fanny rested in the heat of the day at Bolsena;
the prisms are very well defined, though not quite
so neat and regular as those of Staffa, and the
Giant's Causeway, and it is interesting to observe
the transition from this perfectly columnar basalt,
to that which is more irregularly concretionary, and
from this again into the massive and shapeless.
This volcanic country is very richly wooded; it is
rare to see in Italy, at least where I have been, such
extensive woods or such fine trees. They are almost
entirely deciduous—Oak and Chesnut—whereas the
limestone mountains, about Spoleto, Terni and
Narni, are clothed with Ilex, Pine, and Box. I
have not seen a beech tree since I entered Italy;
Virgils "patulæ subtegmine *fagi*" is certainly an
image derived from the skirts of the Alps, and not
from central or Southern Italy.

At that very picturesque place, Aquapendente,
we find ourselves on the northern boundary of the
volcanic formation, the steep hill on which that
town stands, is crowned with a long range of
basaltic rocks, and great masses of basalt lie thick
on each side of the road, in the beautiful but
formidable, steep descent to the Paglia; but the
banks of that river at the bottom of the hill are
of tertiary marl. This marl constitutes the whole of
the third region, from Radicofani to Siena, which is

1848. a strange sort of country, a succession of bare, round, lumpish, monotonous hills, partly cultivated, but very bare of trees, and generally stripped and cut and torn in a strange manner by the action of water.

They reminded me of Carlyle's comparison of the hills about Naseby to the "waves of an indolent sea." The mountain of Radicofani itself is one of these waves, swollen to an enormous size, and capped by an insulated mass of basalt forming a peculiar peak, which is visible from a great distance,—the whole rising to 2500 feet above the sea. This clay country is very generally abused by travellers as excessively dreary and ugly, and in winter it probably is so. Some of the hills have cappings of shelly sandstone, very like the Suffolk *crag* in general appearance; and on these, towns are built,—Siena among the rest. Finding that Susan was already on her way to England, we stayed three days at Siena, just long enough to see the pictures well,— and then came on to this place, which is a great favourite with both of us. Although the weather is extremely hot, the air of Florence agrees very much better with us than that of Rome, and we take very pleasant drives in the cool of the evening. We shall not be able to remain here very long, for the turn which affairs may take in the north of Italy, is so uncertain, that we are rather anxious to get to Turin, to have time to examine the collections there, and then get into Switzerland before anything decisive happens in Lombardy, or the French march across the Alps. At Florence, indeed we might stay as long as we pleased with perfect safety, but 1848. it would be disagreeable to have our retreat cut off, or be obliged to return by sea. At Turin I hope to see Professor Sismonda, if he has not, like so many others been metamorphosed and *bouleversé* by the political and military commotions of the times. Poor Professor Pilla of Pisa, was killed fighting bravely among the volunteers of the university, in the last battle against the Austrians; and another Professor was wounded and taken prisoner. I have not room for more at present.

Believe me ever,

Your affectionate Son-in-law,

C. J. F. BUNBURY.

Florence,

June 26th, 1848.

My dear Mary,

You cannot conceive anything more charming than the country round Florence at this time; it is one rich and luxuriant garden, covered with a thick network of vines, which interlace the trees with the most beautiful festoons of bright green foliage, while under them flourish the richest crops of wheat, maize, flax, clover, and all kinds of corn and vegetable. Multitudes of beautiful wild flowers grow in the fields, and the hedges are over-run with Clematis, Wild Vines, and the climbing White Rose. The country swarms with population, gay villas, hamlets, and single cottages, are scattered so thickly among the vineyards and

1848. orchards for miles and miles around the city, that it really seems (as Ariosto says) as if they sprouted out of the ground. Nowhere, except in the immediate neighbourhood of Rio de Janeiro, have I seen a country so luxuriantly fertile and so smiling.

This rich landscape is enclosed by picturesque hills crowned with convents and villages, and far in the distance to the north-west, one distinguishes the fine outline of the Carrara Mountains. You will find a beautiful and very true description of the environs of Florence in the first volume of Henry Napier's History, at the beginning of the second chapter, and I dare say you know the eloquent picture in Hallam's Literature of Europe, of the view from Lorenzo's villa at Fiesole. It is a delightful place. We are out a great deal, seeing the galleries and so forth, and take pleasant drives into the beautiful country in the cool of the evening; and we often stay out till the bushes are sparkling with innumerable fireflies. These are exceedingly abundant here, and very beautiful. The sunsets on the Arno are splendid. One evening we went to Fiesole, saw the gigantic remains of its Etruscan walls, enjoyed the view from the top of the hill over the Vale of Arno, and thought of Galileo, of Milton, and of Lorenzo de Medici.

Another day we drove to Pratolino, a beautiful villa of the Grand Duke's, six miles off, among the mountains, where the air is delightfully fresh and invigorating; it is like an English park,—like a park in Herefordshire or Shropshire, or some other of the hilly counties of England,—with fine woods

of Oaks and Chesnuts, intermixed with a variety 1848.
of cultivated trees and beautiful green grassy slopes
and glades, and the views of Florence, and the
country below are delightful. We have not been
able to see La Petraja, another famous villa,
because the Court is there at present. I need not
expatiate on the riches of Florence in the way
of paintings, because you will hear all about it
from Susan, and a better account than I could give,
but we have to day seen one thing we did not
see when she was with us,—the painting of Giotto
in the Bargello, including the famous portrait of
Dante, which was discovered under a thick coat
of whitewash a few years ago.

Florence is certainly to me the most agreeable of
the great towns of Italy.

Science, literature, and art are almost neglected
at present in Italy, and whichever way we turn, we
hear and read of nothing but "bombs, drums, guns,
bastions, batteries, bayonets, bullets!" Seriously,
the struggle with Austria absorbs the whole thoughts
and souls of the Italians,—and no wonder, for it is
no light enterprise that they have on their hands.
The fact that so many Professors and other men
of ripe years and peaceful professions have joined
the volunteers, shows how completely they look
upon it as a national and a holy war ; nor does
their enthusiasm seem yet to be at all damp, in
spite of the heavy loss that the Tuscans sustained in
the last battle ; fresh volunteers are continually
setting off for the army. But amidst all this, the
Italians are as childishly fond of festivals and fine

1848. shows as ever. The eleven days that we have been at Florence (I am now writing on the 28th), there have been four great *festas*, with splendid processions —the Grand Duke with all his Court going in procession to the cathedral, the whole of the National Guards under arms and a noise of bells sufficient to give one a Mahometan horror of them for the rest of one's life ; and all this we are to have over again to-morrow. I am utterly sick of their processions and everything of the kind. We had a good sight of the Grand Duke one day at a review ; he is not handsome, but he is a very good man and good Sovereign, and I believe is very popular here.

I conclude you are now at Kinnordy ; pray remember me very kindly to Mr. Lyell.

Is the Linnœa* still flourishing? How much we shall have to talk over when we meet, which I trust will be pretty early in the autumn.

Much love from Fanny.

Ever your very affectionate Brother,

C. J. F. BUNBURY.

---

* Linnœa Borealis,

To Charles Lyell, Esq. of Kinnordy.

Florence,
July 2nd, 1848.

My Dear Sir,

I am happy to say that I have at last been able to meet with one of the two works on Dante, which in your letter of the 18th May, you asked me to procure for you,—namely the "Orologio di Dante Alighieri."

We have seen two interesting records of the great poet, with both of which you are doubtless well acquainted, the one is the Fresco painted by Giotto in the Bargello,—cleared within the last few years from the coat of whitewash under which it had long been concealed, — where Dante appears among a number of magistrates and other notables of Florence, who are offering up their thanksgivings for the victory of Campaldino; so at least the *custode* told us. There is a print from this portrait of Dante, which gives a very correct idea of it, I have no doubt you are acquainted with it. The other is the "mask" or plaster cast taken from his face after death, which is preserved in the Casa Torrigiani, and of which there is an excellent engraving in your book. We found that your name and reputation as a worshipper and illustrator of Dante, were well-known to the Marchese Carlo Torrigiani, the possessor of this treasure. It is highly interesting to see two such authentic records of the features of the great Florentine poet preserved in his native city con-

1848. firming each other, and showing the changes produced by years and misfortunes.

I was much gratified by your approbation of my
book; I hope it will not be the last that I shall be
able to send you. Your letter altogether was very
gratifying to me, although you really made me
ashamed of myself by praising me so much beyond
my deserts.

We were very sorry to lose Susan's company, but
we little dreamed of the dangers and miseries she
was destined to go through

Florence is to my taste, the most agreeable city in
Italy, and the most beautiful that I know in any
country, Venice perhaps ought to be excepted, but
it is so long since I have seen Venice, that I cannot
fairly compare them. The interesting historical
memorials and poetical associations of which
Florence is so full, the innumerable noble and
stately buildings ; above all the matchless cathedral
and the exquisitely beautiful bell-tower, the great
names and noble deeds which these buildings recall
to our minds, the admirable collections in art and
science, the beauty and convenience of the streets,
the civilized and pleasing manners and agreeable
appearance of the people, the rich and smiling
beauty of the surrounding country, the aspect of
order, prosperity and plenty everywhere apparent,
combine to render Florence delightful. The contrast it presents to the Roman State, both in the
flourishing appearance of the country, and in the
character and manners of the people, is very striking.
Another thing which has contributed to render my

stay at Florence this time particularly agreeable, is 1848.
that I have been able to ride my favourite *hobby* of
botany with great satisfaction, having become acquainted with Professor Targioni, who has taken
pains to show me everything here that is worth
seeing in the way of science, and especially the
rich botanic garden and herbarium belonging to the
Grand Duke. These are under the care of Professor Parlatore, who is just now absent, but I have
had every facility for examining the herbarium,
which is rich in Brazilian and Cape plants, especially
in the Ferns of Brazil, collected and described by
Raddi.

Professor Targioni is grandson of the celebrated
botanist of that name, the friend of Micheli, and
after whom the *Targionia* was named. His collections
and those of Micheli, are now incorporated with
the Grand Duke's herbarium, having been bought
from the present Professor. I have not failed to pay
my respects to the monuments of these two
botanists, and of Raddi, in that noble Church,
Santa Croce. The Grand Duke, in addition to
his other merits, is an enlightened and liberal
patron of science, and has shown marked attention
to distinguished scientific men of other nations who
have visited Florence. He is excellent, both as a
man and a Sovereign, and is deservedly popular ;
he has, however, a much easier game to play than
the good Pope, whose position is very peculiar and
embarrassing, and has to deal with a people far less
civilized, less rational, and less steady than the

1848. Tuscans, and with the consequences of the mis-
government of his predecessors.

The future of Italy appears cloudy and uncertain,
the result of the war against Austria constitutes
only a part of the question; for supposing the
independence of Italy secured, it will then remain to
be seen how far the people are qualified to make a
wise and good use of their freedom.   I have good
hopes for Tuscany and Lombardy and Piedmont,
but I have no faith in the Neapolitans, and little in
the Romans.   It is satisfactory, however, to see
that the Italians in general are well on their guard
against France, and rather warned than seduced by
her example.   Many secret attempts have been
made by French emissaries to republicanize Italy,
but happily with almost as little effect as in
England.

I hope you have had a good share at Kinnordy
of the fine weather of which I hear such flattering
accounts from England, and that it has been
favourable to your health, and has enabled you to
enjoy thoroughly the visit of your sons and daughters.   I have a lively (and I assure you, a most
agreeable) recollection of Kinnordy, and can
imagine the whole party assembled there.

My Wife sends kind love to you and all your
family.

Believe me ever,

Yours very sincerely,

C. J. F. BUNBURY.

My Dear Father,

There is no end to our adventures and mis-adventures, I never met with so many in any previous journey, no not in Brazil, nor the Cape. We left Florence on the 5th, by vetturino, intending to make the journey to Genoa in five days. Towards evening, when we had passed Pescia, and were about eight miles from this place, our vetturino was taken very ill, so that he declared he could not drive any farther, and he asked a man of his acquaintance, whom he met on the road, to take his place, and to procure a cart to convey him to Lucca. After some discussion, this exchange was effected, the vetturino, who really seemed to be ill, went on in the cart, and his friend mounted one of the horses attached to our carriage, but being unused to driving, did not venture to go out of a foot's pace, thus we went on with such extreme slowness, that we were nearly three hours in making out the eight miles, and did not arrive at Lucca till eleven at night. After arriving at the hotel, we discovered that our box of books and a box belonging to Fanny's maid, had been stolen from behind the carriage, though attached by a good iron chain and by stout straps, the chain had been broken and the straps cut. A gentlemen who was present, when we found out our loss, directed us to the captain of the carabinieri of this city, who exerted himself actively on our behalf, and with so much effect, that the next morning the boxes were found and recovered ; but they are now under seal

1848. at the Prefecture here, and we have not yet been allowed to examine them. From what we hear, it would seem that our books are all, or nearly all safe, but the ornaments and small articles of value belonging to the maid are, I fear, irrecoverably lost. The book box was found in a ditch, four or five miles from hence; the other box was emptied, and the clothes, etc., scattered about the fields. The place where the robbery was probably committed was somewhere about the former frontier between Lucca and Tuscany. Nothing positive has been ascertained about the plunderer, our suspicions fell on a countryman with a cart, who accompanied us for some distance in the dusk of the evening, sometimes walking by the side of our horses and conversing with the driver, sometimes walking along side the carriage, and talking to us in English, for he said he had spent some years in England,— but always keeping his cart behind our carriage. The magistrates also seem to suspect this man, but nothing very positive appears against him, and no steps have been taken to secure his attendance. We have had an opportunity, which we by no means wished for, of becoming acquainted with the formalities of justice in this country. Nothing could be better than the conduct of the police, and everybody has been very civil, but the mode of taking our depositions was exceedingly tedious; we are in hopes of getting back our property in a few days, but we do not exactly know when. The worst of it is, that being hampered by our agreement with the vetturino, we are obliged to keep him and his horses all this while idle at Lucca, and we cannot 1848. go to the Bagni or make any excursions, from the uncertainty when we may be wanted at the police office. The landlord of our hotel thinks the theft was committed by some of those men who have deserted, or been sent away in disgrace from the army. It is certainly not a good time to be in Italy, and we begin to wish ourselves fairly at home. Not that I suppose the Italians (or at any rate the Tuscans) to be on the whole worse than any other people, but the state of the country at present is such as to give great facilities and advantages to the ill-disposed.

(July 9th.) We have seen our property, but we do not yet know when it will be restored to us. All our books are safe, the thieves having no doubt voted them useless lumber, but two Roman mosaics of some value, which were in the same box, are gone, and poor Elizabeth has lost everything of value that she had. The police give us very little hope of recovering any of these things. In all probability the thieves will have carried them off to Leghorn, and disposed of them to the Jews, and other bad characters, who swarm in that sea-port. We are in hopes of having our books restored to us the day after to-morrow, and of being able then to continue our journey, but this depends upon the will and pleasure of the tribunal. The delay is very wearisome, but our consolation must be, that the affair might have ended worse.—Our landlady gives a very bad character of the country people hereabouts.

1848.   A shocking thing happened last night a mile or
two outside the city: some huts, full of the farming
stock and produce of the harvest, belonging to two
honest poor families, were set on fire and totally
destroyed, the perpetrator, a notoriously bad
character, who had been often in prison, is well
known, and it is known that he committed this
abominable act in revenge for having been detected
in a theft by the poor people. He is not yet
taken.

(*Genoa, July 17th*). Our books and boxes were
restored to us on the 11th, and to the credit of
Tuscan justice be it said, we had not a farthing to
pay in fees or law expenses. The magistrates
behaved to us, all through the affair, not only with
civility, but with the greatest possible good nature,
patience and good humour. Still, the circumstances
of our detention at Lucca were not such as to leave
us an agreeable impression of the place, especially
as we were uncomfortably lodged in the hotel, yet it
is a place of some interest, the cathedral is noble,
and we particularly admired the pictures of Fra
Bartolomeo, one in the Cathedral and two in the
Dominican Church.

The views from the ramparts too, are very pretty.
We got away from Lucca at last in the afternoon of
the 11th, and slept that night at Pietra Santa, and
the next at La Spezia. We were tempted to
stay two days at that charming place Sestri di
Levante, and very much I enjoyed them ; the
situation of the Inn there is delightful, and the
tranquillity, the beauty of the scenery, the freedom

from noise and from insects, the fresh sea air, the sea 184
bathing, and quiet rambles among the Pines on the
little peninsula, were most refreshing and restorative
after all the heat and worry we had gone through.
Yesterday we arrived here.

I have filled this letter with nothing but our
personal misadventures, but I hope in the next to
give you some more general news.   Believe me,

Ever your very affectionate Son,

C. J. F. BUNBURY.

Turin,
July 28th, 1848.

My dear Father,

We are here at a very critical and anxious
moment; it is known that for *four* days the
Piedmontese army under Charles Albert has been
engaged in a serious battle with the Austrians,—
that it has fought at a disadvantage and has been
compelled to retire to the right bank of the Mincio;
but the final result is yet doubtful.   It is known
that the number of killed and wounded is great—
6,000, it is said, in one day's fighting,—but no
particulars are known.   Almost every one at Turin
has relations or friends engaged, and you may
conceive the painful excitement and suspense that
prevails.   It is said that the Austrians have 120,000
men, while the Italians (including volunteers and all)
have not more than 60,000, and these moreover at
the beginning of the fight were spread over a great
extent of country from Rivoli down to Gover-

1848. nioli, near the junction of the Mincio with the
Po. Whether it was wise to occupy so extensive a
line, is a question which you as a soldier can answer
better than I can. The volunteers of Lombardy,
10,000 strong, who were on the extreme left in
the valley of the Adige above Verona, ran away,
and this compelled the rest of the army also to fall
back, lest their left flank should be turned; a
Genoese regiment is said to have also run away,
and the Modenese went over to the enemy! I
can well understand the bitterness of feeling
which the Piedmontese betray when they see
the other Italians, for whose freedom they are
lavishing their blood and their resources, desert-
ing them in the hour of danger. The Piedmontese really seem to be the only Italians who
have the good sound durable material of soldiers in
them; the rest, even the Lombards, showed prodigious ardour at first, but

> "Their valour like light straw on fire,
> A fierce but fading flame!"

It appears that the King wished to make peace
with Austria on the terms you mention, fixing the
Adige for the boundary, but the vehement opposition
of the Provisional Government of Milan prevented
the conclusion of the treaty; they would have
Venice too; and then when it comes to actual
fighting, the Milanese are the first to run away.

We are fortunate in having found a very agreeable
and useful friend here in M. Prandi, who was one
of the political martyrs of 1821, and an old friend

of Fanny's, having known her and her family for 1848.
twenty years. He is very lately returned from
England, where he lived so long, that he not only
speaks English perfectly, but has very much
of the manners and taste of our countrymen, and he
is now a member of the Chamber of Deputies here.
He is an uncommonly agreeable man, full of information and intelligence, and I think of remarkably
sound and rational views on political matters.
Fanny indeed is *en pays de connaisance* at Turin, for
M. de Marchi, another of the Deputies, and likewise
an exile, in consequence of the affair of 1821, was
intimately acquainted with her family at Edinburgh.
M. Prandi is much disheartened and out of spirits
at the state of affairs in this country : he thinks the
Chambers are going on ill, that there is deplorable
want of method and organization and management
—no parties—everybody acting according to his
own fancies ; and that they waste time in loose and
rambling debates upon trifles, neglecting to provide
for the urgent wants of the crisis—for instance,
they have just now consumed two days' sittings in
discussing a proposition for demolishing the forts at
Genoa ; rather ill-timed surely, in the middle of
a dangerous war ; however, it has been carried that
the Castelletto and the Fort St. George should
be demolished, and a commission is appointed to
carry this into effect. Some wanted to demolish
also the citadels of Turin and Casale.

(July 29th). Yesterday we went to the Chamber
of Deputies, M. Prandi having given us tickets for
the Ambassador's box, so that we saw and heard

1848. very well.   It was a sitting of some interest and
excitement, both on account of the painful excitement respecting the army, and because the new
Ministers were expected to take their seats for the
first time.

I should have told you that, in addition to the
military crisis, we are here in the crisis of a Ministerial Revolution.   The late Ministry indeed, of
which Count Balbo was the head, resigned nearly a
month ago, but had ever since continued in their
offices *ad interim* ; because no successors could be
found to them.   At length M. Collegno who was
charged to form a Ministry, has succeeded in
forming one, which our friend M. Prandi thinks
very badly chosen.   It is certainly strange, considering the present importance of Piedmont, and
the leading part it plays, that Collegno himself is
the only Piedmontese in this new Ministry, which is
made up of men from all the different divisions of
Northern Italy.   It is very unlikely that they will
pull well together.   Collegno lived many years in
exile at Paris, and has considerable reputation as a
geologist, but does not seem to be so well qualified
as a Minister.   There is another geologist in the
Ministry, the Marquis Pareto, who was in the late
Ministry too, but is said to be a rash enthusiast in
politics.

Well, the sitting of the Chambers yesterday, was
quieter than I had expected, but we heard a great
many of the members speak, and most of the
principal speakers, in particular, Brofferio, who has
a great reputation as an orator, and who certainly

has great energy and an impressive manner; but I 1848. could not very well follow what was said. The Deputies speak from their places, not from a *tribune* as in France; it struck me that they spoke more briefly than our Members of Parliament generally do, and several of them in a more conversational tone and manner; but some seemed to read their speeches. The same Members spoke again and again, seemingly without any restriction, in the course of the same discussion. What struck me most was the very disorderly conduct of the audience in the public gallery, who three different times interrupted the business of the house by tumultuous outcries, and were only called to order by the President—not turned out as they ought to have been.

This morning, we hear the crowd around the Chamber of Deputies was very menacing, and three working men actually forced their way into the Chamber itself—an attempt to renew the scenes of Paris. At last the National Guards cleared a wide space before the building, and guarded it with fixed bayonets, admitting none but those who had business or a right to enter.

(July 30th). We learn that the Chamber of Deputies yesterday passed a law investing the King with dictatorial power, legislative and executive, for the duration of the war, and to day though it is Sunday, the peers held an extraordinary sitting, and voted the same law. This we have from M. Moris, the Professor of Botany, who is one of the peers. It is the wisest thing they could do in such a crisis, when the cavils and crotchets and uncertainties of an

1848. inexperienced deliberative assembly might seriously impede the great work of national defence. The King has published a proclamation, announcing that the Austrians had proposed a suspension of hostilities, but on such terms that he could not consent even to discuss it. There seems to be no more certain intelligence from the camp, but it is supposed that the fighting has ceased, both parties being tired out. All the accounts of the action that have yet appeared, are very indistinct; all that I can make out is, that the left wing of the Piedmontese army was thoroughly beaten; that the Austrians crossed the Mincio between Peschiera and Goito, but that the rest of the army retired in order to the position of Goito and was concentrated there.

The celebrated Gioberti has joined the new Ministry, but without any special office—*sans porte feuille*. It is astonishing to what a degree this man is idolized in Italy. We travelled in his track falling in with him at Rome, at Florence, at Lucca, at Carara and at Genoa; and everywhere he was received with an enthusiasm hardly to be described. Wherever he came the streets were hung with banners and strewed with flowers, "*Viva Gioberti!*" placarded everywhere; a guard of honour placed at his door; deputations of the principal inhabitants waited on him to invite him to their Clubs and Societies, and it was publicly announced that Vincenzo Gioberti would on such a day and hour, honour such or such a club or Academy with his presence. He is certainly a man of great eloquence, and some of his writings are much admired by the cleverest Italians 1848. whose opinions we have had an opportunity of hearing, such as Dr. Pantaleone, but his writings are not of a kind adapted "*ad captandum vulgis*," and his amazing popularity seems to be owing partly to his attacks on the Jesuits, and mainly M. Prandi says to his setting himself up as the champion of the lower priesthood, representing the union of the sacerdotal and democratic elements.

The prevalence of the ecclesiastic and priestly spirit in the present movement in Italy, is one of the most remarkable phenomena, and peculiarly distinguishes it from all the different Revolutions in France. I believe that the influence of the priesthood, and the zeal of the Roman Catholic Church, among high as well as low, so far from diminishing with the advance of political freedom, are, if possible stronger now than ever. The two most popular and influential writers, Gioberti and Count Balbo, show in their writings (at least so I am told, for I have not read them) Church and State zeal worthy of Mr. Gladstone in his early days, and dwell upon the unity of religion and the power and influence of the Pope, as the great securities for the national unity and independence and future ascendancy of Italy. Fanny is reading the works of both these authors, which I really have not time to do, so I take my account of them from her. We have been introduced to Count Balbo (the late Prime Minister) by M. Prandi, and were at his house one evening; he and his family made a very agreeable impression on us. I have likewise

1848. received a great deal of civility and kindness from
the scientific men here. Professor Sismonda and
his brother, and Professor Moris, the latter has
made me a really magnificent present of his great
work on the plants of Sardinia, and I have had
great satisfaction in examining the collection of
minerals and fossils in the museum.

The minerals of the Piedmontese Alps are
most beautiful. Thus though Turin is generally
reckoned a dull place, we have found a great deal
of interest in it at this time, and should have
passed our time very pleasantly if the weather had
not been so tremendously hot. To-day indeed, it
is comparatively cool and pleasant, in consequence
of a heavy thunderstorm last night.

We have decided on going to Geneva by the
Mont Cenis, in consequence of M. Moris's information, for he says it is the best of all the
passes for botany, and that we ought to go before
the grass is cut on the heights. But we talk of
going from Geneva to Milan by the Simplon, and
returning again into Switzerland by the St.
Gothard. I will write to you from Geneva.

I should have thanked you sooner for your
letter of the 3rd, which I received on our arrival
here.

What a frightful state of things in Ireland! I
expect every day to read of the rebellion having
broken out, for one cannot conceive that so many
months of preparation and boasting, will lead to
nothing. I am glad the Government measure for
the suspension of the Habeas Corpus was passed

so rapidly, and with such a general consent of parties in both houses; it is no doubt the wisest plan that could be adopted, though in one point of view it will be an advantage to the rascally leaders, O'Brien and others, who will be safe in prison while the mortal struggle takes place. It is a sad and painful time for all who have friends in Ireland, I am very happy that Henry is clear of it.

(*July* 31*st.*) Our friends here are very much out of spirits about public affairs, which indeed have an aspect sufficiently disheartening. The Piedmontese troops, it is true, fought gallantly, though forced by the vast superiority of numbers of the enemy, and by the misconduct of their allies, to retreat, they did so in perfect order without losing a single gun or trophy of any kind, and they took 2,000 prisoners. But it does not seem likely, that with all their courage, they will be able to cope with such overwhelming forces as Austria can bring into the field ; their own countrymen admit that they have no generals, and the people of Lombardy who are so much more deeply concerned in the issue of the war than any other, do nothing.

The Provisional Government of Milan, engaged to feed the army, and have managed it so ill, that the troops have always been in want, and during these last battles, were two days without any food at all. It is no wonder the Piedmontese are both angry and disheartened. It is sad after all the scenes of rejoicing and exultation and enthusiasm, that we witnessed during our first two months in Italy, to see the fading away of such bright hopes,

1848. and the gloom that is spreading over all.—It is said that one of the new Ministers is gone to Paris, to ask for French assistance; but I hope that neither the King nor the people of Piedmont will submit to this.   It is a move of the Republicans.—Mr. Abercromby is gone to the King's head quarters.   It is a most anxious time.

Ever your very affectionate Son,
C. J. F. BUNBURY.

Geneva.
August 8th, 1848.

My Dear Mr. Horner,

I had no time to write to you from Turin, as I had intended to do, being very busy all the time we were there.   It was a time of great and anxious interest in respect of politics, but I know Fanny has written you a very full and very correct account of all that we heard concerning politics and war, so I will stick to geology.

I went to inquire for Professor Sismonda the morning after our arrival at Turin, and met with a very polite and kind reception from his brother, M. Eugene Sismonda, who was delighted when he heard that I was brother-in-law to Lyell, and your son-in-law, and spoke warmly of the attentions he had received from you on his visit to London, when you were President of the Geological Society.   The Professor returned three days after, and welcomed me very cordially, so you see my visit was fortunately timed, in that respect at least.   Both

brothers were as friendly and obliging as possible, 1848.
though of course the Professor, immediately on his
return from so long an absence, and that on an
important Government mission, had no time to
spare, and I could not see much of him. Every
facility and assistance were afforded me for
studying the fossil plants of the Tarentaise, and I
think I have worked out the botanical part of the
question pretty completely. But I found great
difficulty in the investigation, and was obliged to
proceed with much caution, for the plants are in a
very perplexing state. They are jumbled and
mashed up together in still greater confusion than
in our coal shales, generally much broken, and all
converted into white talc ; the venation in general
very indistinct, though here and there it is remarkably well preserved ; and what is most puzzling,
many of them seem to be distorted and disfigured
in an extraordinary way, I suppose by the crystal-
lizing process, which converted the shales into
talcose slates. I was really bewildered at first
by the variety of forms of leaflets which I saw in the
same frond, and even on the same division of a
frond ; but by degrees I perceived that this must be
occasioned by the metamorphic and crystalline
action which the schists and the fossils contained in
them had undergone—just as Daniel Sharpe has
shown that the shells in the slate rocks of Wales
have been distorted by the same cause. This of
course renders the determination of the species
difficult and rather hazardous, but I have nevertheless made out a certain number to my satisfaction

E E

1848. and these are all truly carboniferous; so as far as I
go, I entirely confirm Brongniart's conclusions.

But the variety is less than I expected; in the Turin
collection, which is said to be the most complete,
I could find only fourteen different *forms* (I will
not venture to call them *species)* of which nine are
Ferns, two Calamites, and three Asterophyllites.
Those that I can determine as truly coal plants, are
*Neuropteris tenuifolia, Odontopteris brardii, Pecopteris
arborescens, Calamites opproximatus,* and *Asterophyllites
equisetiformis* (of Lindley and Hutton). Three other
Ferns appear to be identical with coal measure
species, but the specimens are too imperfect to
allow me to speak with certainty ; and to differ a
little but hardly specifically from plants found in
the coal formation.

I have given you these details because you
brought this extraordinary anomaly prominently
before the notice of the G. S. in your first anniversary address, and as I remember, took a great
deal of interest in the question. I have made
copious descriptive notes from M. Sismonda's
specimens, and have materials for a short paper on
the botanical part of the subject, but before I draw
it up I must examine what has been previously
done, and especially Adolphe Brongniart's paper on
the plants collected by Eliede Beaumont. As to
the geological part of the question, it appears to me
a problem at present insoluble. Sismonda is not
only himself satisfied that the Ferns are contemporaneous with the Belemnites, but he told me
that Murchison, who visited Petit Cœur this summer

came to the same conclusion. Now I know that 1848. Murchison when I saw him at Rome was completely sceptical on the subject, so that the evidence must be very strong to have satisfied him. The Abbé Chamousset too (whom I saw at Chambery) says that no one who has been at all accustomed to the investigation of Alpine geology, can doubt, after examining the localities in the Tarentaise, that the strata containing Belemnites do really form parts of the same formation, and are contemporaneous with those containing coal-plants, though he differs from Sismonda and E. de Beaumont in regard to certain other strata containing *Ammonites*, which they consider as contemporaneous with the anthracites, while he thinks they are of somewhat later date.

With all this weight of evidence, I think we can hardly dispute the fact of the co-existence of the Belemnites and the coal-plants ; and to whichever we give the preference in determining the epoch of the formation, the fact stands as a local phenomenon very difficult to explain.

The collection of minerals at Turin is exceedingly rich and beautiful, and admirably arranged—better than any other public collection I ever saw. The series of minerals from the Piedmontese Alps most splendid, especially the crystallized magnetic iron pyrites, and carbonate of iron from the mines of Traversella, the augite from the same locality, the idocrase from La Mussa, and the garnets and diopside from the valley of Ala.

The fossils also appear to be a very fine collection and beautifully arranged, but of these I was not so

1848. well able to judge. Dr. Sismonda (Eugene) seems
to have a particular talent for arrangement ,and he
is also a very active and zealous palœontologist. He
gave me some of his valuable memoirs on Piedmontese fossils. I was much pleased with both the
Sismondas ; indeed we found our stay at Turin very
pleasant on account of the agreeable acquaintances
we had there ; I liked M. Prandi exceedingly, and
by him we were kept up to the current of the
political events that were taking place, and indeed
admitted in some measure behind the curtain. He
and all our friends were naturally very anxious and
uneasy about the state of their country ; dissatisfied in
the highest degree with the new Ministry, blaming the
conduct of the Chamber, which frittered away their
time in endless and useless cavillings ; indignant
at the supineness of the conduct of the other
Italians, seeing their country engaged almost single
handed against the immensly superior power of
Austria, and dreading the intervention of France
hardly less than the success of the Austrians—they
might well feel uneasy if not gloomy.

The prospect is indeed very uncomfortable and
very unlike that which existed when we entered
Italy. Fortunately England has a worthy representative at Turin, in the person of Mr. Abercromby,
and I hope he may be able to do some good. The
disturbed state of affairs in Italy as well as in
France is sadly injurious to the progress of science
and of all other intellectual pursuits, and I fear
things show little prospect of amendment. Those
men of science who have been drawn away from

their own pursuits to posts of political importance 1848.
have little reason to congratulate themselves on the
transformation. Collegno and Pareto, both of them
geologists of some distinction, make wretched
Ministers. We had a pleasant journey from Turin
hither, and spent twenty-four hours on the top of
Mont Cenis, where I reaped a fine harvest of Alpine
flowers.

Here we have found a plentiful crop of delightful
letters, with most comfortable and satisfactory
accounts of you all. We hope to be in England in
September or early in October.

Ever your affectionate Son-in-law,
C. J. F. BUNBURY.

Geneva,
August 9th, 1848.

My Dear Lyell,

I received on our arrival here, the day
before yesterday, your very agreeable and entertaining letter of the 2nd, and thank you heartily for it.
It was a great treat to find such a budget of letters
waiting for us as we did here, and all full of
comfortable and satisfactory information, respecting
those in whom we were interested. I was very glad
to hear of Joseph Hooker.—Fanny has just been
made very unhappy by hearing (from Mr. Prevost)
of the capture of Milan by the Austrians. I am very
sorry for it too, but she feels it like a personal
misfortune ; she has become so deeply interested in
the Italian cause. It is indeed a miserable con-

1848. clusion, and there must be miserable mismanagement, if not treason.   What a contrast to the
universal enthusiasm and exultation, the wild hopes
and sanguine confidence that we witnessed at every
step of our journey in February and March!
The want of union and of perseverance has
again ruined the hopes of the Italians, and thrown
away the fairest opportunity that has ever been
opened to them.   It is a melancholy business.
And what is to follow?   Mr. Prevost says it is
reported that quarters are already preparing in
Savoy for the French army which is expected; so
Italy is to be again, as it has so often been, the
battle-field of France and Austria;—and will not
the other nations of Europe become involved in the
quarrel?

Well, I will turn to a more agreeable subject.
We enjoyed very much our stay at Turin, and
received all possible civility and attention from both
the Sismondas (for the Professor arrived while we
were there) and also from Professor Moris, a celebrated botanist, to whom I had a letter from M.
Targioni.   I made good use (I hope) of my time
there, worked hard at the fossil plants of the
Tarentaise, and made copious descriptive notes, to
serve for the foundation of a paper on the botanical
part of the question, which I intend to write for the
G. S. as soon as I get home.   As we hope (barring
accidents) to be in England by September or early
in October, I expect to be able to get this ready in
time for one of the first meetings of the session.
The result of my examination (which I can safely

say was careful and laborious) is quite in accordance 1848. with Brongniart's conclusions.

As to the geological question, both Sismonda and' the Abbé Chamousset are quite confident that the Belemnites (which by the way are undeterminable as to species) are truly contemporaneous with the coal plants. There is thus at any rate a hard nut for you geologists to crack, though I should think the difficulty is not so insuperable for you as for those who hold the doctrine of complete and sudden changes at the close of each geological period.

Professor Moris made me a really magnificent present of his great work on the Flora of Sardinia, still in progress, and published at the King's expense; it is the best book that has been written on the botany of the Mediterranean region. He says the present King is a liberal patron of science, and the first of his house that has been so.

We had a pleasant journey from Turin hither, and spent twenty-four hours on the top of the Cenis Pass, at the posthouse by the lake, for the sake of botany. The inn is dirty and dear, but for my part I thought its faults far more than compensated by the profusion of beautiful Alpine plants which I met with; the pastures and interstices of the rocks were perfectly enamelled with them; I do not think I ever made a better harvest. I was quite surprised at the luxuriance of the herbage in the moist meadows around the lake, at 6,700 feet or thereabouts above the sea level, the grass seemed as tall and rich as in the low meadows in England.

1848. Polygonum viviparum, **Gnaphalium dioicum**, and
Arbutus Uva-ursa, which were in great abundance,
brought Kinnordy agreeably to my mind, and
Alchemilla alpina and the yellow Saxifrage put me
in mind of Clova.   Fanny recognized many of her
Auvergne friends, particularly Gentiana lutea, a
magnificent plant.   The Campanulas and two
species of Pink, were remarkably beautiful.   I found
no mosses or lichens, and only one Fern, the
Aspidium Lonchitis.   I had no time or attention to
spare for anything but botany while we stayed on
the mountain, but I should think that an entomolo-
gist might have fine sport there, the insects
appeared very numerous and peculiar, and I saw
many very beautiful butterflies, in particular a large
white one (Apollo if I am not mistaken) with black
and red spots.

The rocks around the lake on Mont Cenis (I do
not know whether you remember them) are of a
curious formation of pure white gypsum, unstratified
and eaten into most fantastic shapes by the weather;
Sismonda thinks they were originally limestone,
metamorphosed by the action of acid vapours at
the time of the eruption of the serpentine rocks
which are found in the surrounding mountains.
He considers the *plain* of Mont Cenis, which contains the lake, as a real "crater of elevation."—We
spent the greater part of a day at Chambery, where
I saw the Abbé Chamousset and a M. Pillet, a
young lawyer, who is a very zealous geologist, and
has the charge of the little collection there ; and we
stayed all Sunday at Aix-les-Bains, a very pretty

place. We have been received here in a most 1848.
cordial and friendly way by the Marcets and
Mr. Prevost; we have spent two pleasant evenings
with the former, and I have become acquainted
with M. Pictet, but have not yet been able to
see Alphonse de Candolle. The botanic garden
here is in beautiful order, but not large,—the
museum apparently much inferior to that of Turin.
We are to dine with Mr. Prevost on the 11th, and
intend to set off the next day for Chamounix,
an excursion to which I look forward with very
great pleasure.

Our further plans are very much unsettled by the
evil turn which affairs have taken in Italy : Milan
is now out of the question, even the shores of the
Lago Maggiore may become the scene of war,—and
I am alarmed about the Rhine, as the German
States, from a misdirected national feeling, seem to
be taking up the cause of Austria, and may determine to support her against France. I hope the
Rhine will not be closed against travellers. The
disaster at Milan seems to have been owing chiefly
to the mad Republicans, who got up a tumult
against the King, and even fired upon him, so that
finding he had to contend against enemies within
the city as well as outside, he gave up the game
and endeavoured to provide for the safety of his
own army and dominions. But it is thought that
Radetsky will march upon Turin, and I fear there is
disunion even there. I think that true and wise
Republicans, like some of your friends in America,
must look with contempt upon such Republicans as

1848. those of Italy. The Lombards have on the whole behaved so miserably, that it considerably lessens one's pity for them ; yet it is sad to think how many of the best of them must again become exiles, and probably for life. I hope that England and France will not allow the Austrians to take any part of Piedmont.

August 10th.—M. Pictet has to-day shewn me some fossil plants of the Alpine anthracite formation from Servoz and the Valorsine and the mountains opposite Martigny. They are some of the same species as those of Petit Cœur, but in addition there is our old friend Neuropteris flexuosa,— *exactly* the same as some of the specimens from Pottsville. This I did not see in the collection at Turin.

We expect to spend all this month in Switzerland. With much love from Fanny to you, and from both of us to Mary,

Ever yours affectionately,

C. J. F. BUNBURY.

Geneva,
August 21st, 1848.

My Dear Father,

We are just returned from a delightful expedition to Chamounix, having spent five whole days there with great enjoyment, though not without experiencing some disappointment from the weather. But before I come to the details of our excursions,

I must thank you for your kind and affectionate 1848.
letter of the 17th of last month, which had been
waiting here for me some time before our
arrival.

Well, now for Chamounix. We started from
this place on the 12th, slept at St. Martin, and
reached Chamounix (after a severe jolting in a char)
in the middle of the next day. Both these days
were extremely fine; we saw Mont Blanc and his
attendant Aiguilles in high beauty, and reckoned
confidently on a fine day for the Montanvert, but
the wind changed in the night, and the morning of
the 14th was wet and dull. About the middle of
the day it cleared up, and after an early dinner we
set out for the Montanvert, with a good guide,
Fanny on a mule, and I on foot. Unluckily the
clouds gathered again thickly before we reached the
top, and the Mer de Glace looked gloomy and
awful under the heavy masses of dusky vapour that
hid the peaks of the Aiguilles around it. We
collected many interesting Alpine plants, but the
night came on almost before we could reach the
valley, and we returned to our inn in the dark, and
in a thunderstorm, which was more romantic than
agreeable. The weather was more favourable to us
on the 15th. We went to the Glacier de Bossoms,
and crossed over it under good guidance, and
with crampons on our feet, as owing to the coolness
of the weather, the ice was uncommonly hard and
slippery. We got across quite safely, and returned
through a very pleasant path through the fir
woods, crossing several torrents on the sides of

1848. which a beautiful Epilobium was blossoming profusely among the loose rolled masses of granite.
The path led us at last to the Cascade des Pelerins,
a very pretty and singular waterfall, which I did
not see in our former visit to Chamounix in 1829 ;
indeed I believe it is one of the new *lions* of the
place. The peculiarity of it is, that the stream
falling perpendicularly into a basin or hole in the
rock, is thrown up again with great force, and in a
multiplicity of beautiful curves, like a *girandola*.
We did not see it in perfection however, as the
season has been a very cool one, and the stream
consequently much smaller than it is in hot
summers. The next day we ascended to the Croix
de la Flégère, but we were again unlucky in the
weather, it was sunny and fine when we set out,
but clouds came up the valley before we reached
the top, and soon enveloped us ; we gained the
pavillion of La Flégère in time to take shelter from
a pelting shower, and stayed some time botanizing
in the Alpine pastures around it, — the clouds
opening from time to time, and giving us magnificent
glimpses of the glaciers and the Aiguilles. In
descending we were caught in a furious storm, and
returned to Chamounix drenched to the skin. The
following day was so wet that we did not go out
at all, but the 18th was beautiful, and we made a
delightful expedition to the Col de Balme,—the
most interesting I think, of all the excursions around
Chamounix. The view from the Col was most glorious.
Mont Blanc and the whole range of Aiguilles
displayed in all their sublime majesty, in greater

splendour than from any other point that I know; 1848.
The Arve winding and glittering through the valley ;
the Aiguilles Rouges, magnificent pinnacles of rock,
towering up on our right hand, and behind them the
snowy crest and dark cliffs of the Buet. Then looking
another way, we saw the peaks above St. Maurice, and
the whole range of Alps along the north side of the
Valais, terminated by the Blumlis Alps, and the
Jung Frau herself, with their snowy peaks, perfectly
clear and well defined against the sky, above a long
belt of clouds, which stretched along their flanks.
It was certainly one of the most splendid scenes
I ever beheld in my life. We did not descend from
the Col by the usual route but made a circuit
through the Alpine pastures between it and the
Tête Noire, to a slate quarry, in which fossil Ferns
are sometimes found, and descended from thence by
a very steep path to the village of Tour, where we
had left our char. I was quite in ecstacies and so
was Fanny too, with the variety and beauty of the
little Alpine plants that carpeted the Col and the
mountains about it. I never enjoyed a day more.
We were out nearly ten hours. We quitted Chamounix on the 19th and arrived here yesterday,
both of us, I think, the better for our expedition
though plentifully sunburned.

Italian politics are now rather a sickening subject,
the result has been so different from what we hoped,
and what a very few months ago seemed probable.
It is admitted on all hands that the Piedmontese
fought gallantly, but they were left to maintain
the struggle almost single-handed against all the

1848. power of Austria, and their Generals from the King downwards, seem to have been sadly deficient in military skill and ability. The conduct of the Lombards has been so miserable, that one is almost tempted to say that they deserve their fate, and yet that would be hard, for their imbecility was no doubt produced in great part by the bad Government under which they have so long suffered.

For four months the Milanese territory was free from the presence of the enemy; and in all that time they provided neither money (though many of them are very rich) nor ammunition, nor stores of any kind for an army, nor men capable of fighting. The Piedmontese army was nearly starved in the midst of a friendly country, and *that* the richest and most fertile country in Europe.

It remains to be seen what the mediation of France and England may be able to do for a people who have shown that they can do so little for themselves. I must say that the present Government of France has behaved very well with respect to Italy, showing no rapacity nor undue eagerness to interfere, but a moderate and pacific spirit. I do not think any French Government ever behaved so well before.

General Cavaignac and his coadjutors seem to be doing a great deal of good in that country, and to be really the very men to govern France.

The Irish rebellion has turned out a much less formidable affair than I had expected; it is clear that O'Brien and the other would-be leaders had completely miscalculated their powers of doing mischief, but it is clear also that they wanted only 1848. the power and not the will, to involve the whole country in bloodshed and confusion, and I would not spare them. The apprehension of any general and serious insurrection in Ireland seems to be over, but I fear there may be many isolated acts of ferocity against individual landlords, and it is much more difficult to guard against a system of assassinations, than against an open out-break.

We have met with much kindness and friendly attention from several families here,—the Marcets, Prevosts, and De la Rives,—friends of Fanny's family, and I have become acquainted through M. Marcet, with M. Alphonse de Candolle, son of the great botanist of that name, and himself a botanist of high reputation. The society of Geneva seems to be very agreeable, but we shall not stay here long at this time, for the season is growing late, and we are anxious to see the Oberland and the Lake of Lucerne. We intend to set off on the 24th for Berne, where I shall hope to hear from you, as we shall return thither again after visiting the Oberland.

I do not know whether we shall fall in with Edward, but it is not unlikely. Pray give my love to Emily and Cecilia, and believe me,

Ever your very affectionate Son,
C. J. F. Bunbury.

1848.   Lausanne,
                                      August 26th, 1848.

My dear Mary,

We are halting for a day at this beautiful
place, having been persuaded by Miss Allott to stay
to dine with her, instead of going on to Vevay this
afternoon as we had intended; so between our
return from the Cathedral and the hour when we
are to go down to Ouchy, I take the opportunity of
beginning a letter to you.

We both of us botanized zealously during our
stay at Chamounix, and with great success,
especially on the Col de Balme.   The Alpine flowers
are charming; no words can do justice to the
beauty of the little Gentians which spangle the turf,
and whose flowers are of a blue more exquisitely
vivid than can be imagined.   On that one Pass we
found five species of Gentian in blossom, one of
them (G. purpurea) a fine showy plant, with large
flowers of a very peculiar tint, between purple
and chocolate colour.   The root of this is much
valued as a medicine by the people of those Alps.
I was reminded of the Clova mountains, by finding Azalea procumbens in abundance both on the
Flègére and the Col de Balme, and I thought of
your White Mountain collections when on the top
of the Col de Balme.   I saw that plant intermixed
wita Cetraria nivalis and cucullata.

Altogether on the Mont Cenis and at Chamounix,
we collected and dried at least 70 species of plants.
I shall be curious to see how far the calcareous
Alps of the Bernese Oberland differs in their

vegetation from the granitic Alps of Savoy. I was 1848. particularly pleased with Mr. Frank Marcet and his Wife.

I had great pleasure also in becoming acquainted with M. Alphonse de Candolle, who was as courteous and obliging as possible, and paid me every attention—went through the Botanic Garden with me, gave me specimens of many plants from it, and allowed me to examine at my leisure his herbarium, which contains all the authentic specimens for his father's great work, and is arranged in exact correspondence with that work. He is an agreeable and sensible man, though much less brilliant I understand than his father. I must say that I have been received in a most agreeable manner by all the men of science that I have met with in the course of our tour—at Florence, Turin, Chambery and Geneva. The present Radical Government of Geneva have clipped, most unmercifully, the salaries of the Professors, and the allowances for the Botanic Garden and other scientific institutions. They seem to lean to Ledro Rollin's notion, that learning and science are useless and that the untaught instinct of the masses is sufficient for everything. Your American friends are wiser.

We are still here, at Lausanne. On the 27th, Mr. Haldimand, whom we have found remarkably agreeable, pressed us in so kind and cordial a manner to stay over to day, that we consented, and indeed the scenery is so lovely that one is easily tempted to linger.

1848.   We have a delicious view from the windows of our room in the Gibbon Hotel. We have found in this room a large old bible which belonged to Edward Gibbon, not exactly the most characteristic relic of the great historian.

(Vevey, August 29th). We spent a very pleasant evening with Mr. Haldimand, whom I like paticularly, and came hither yesterday. It was a brilliant day, and hot enough for Italy, and the scenery seemed to grow more lovely as we came on towards this end of the lake. We are here in a first-rate hotel in a most delightful situation close to the lake; this day too, is glowing, and there is that delicate blue haze over everything which belongs to hot climates. To my mind the scenery of Vevey is the most charming that we have beheld since we have been travelling together. Yesterday evening we drove in a car to the castle of Chillon, saw the famous dungeon of Bonnivard, the torture chambers and other curiosities of the place, and enjoyed delightful views. To-day we propose a boating excursion. Mr. Haldimand lent me the last " Edinburgh Review," and I have been reading the article on Goldsmith, which is very entertaining.

(August 30th). We made a very pleasant boating excursion yesterday evening, crossing the lake to St. Gingolph and thence to the mouth of the Rhone. The evening was lovely, the lake as smooth as glass, and nothing could be more softly beautiful than the reflections of the mountains in the water, or than the tints of its surface at sunset.

The entrance of the Rhone into the lake is a 1848. curious sight, which I never saw before. The strong contrast of the muddy, whitish river, and the transparent blue water of the lake, and the way in which the mud is seen spreading in clouds through the clear water, and gradually sinking, is very striking. I remember it is particularly mentioned in Lyell's " Principles of Geology."

It is difficult to leave this charming country, but I suppose we shall get to Berne some time or other, especially as we expect letters there.

Love from both of us to your Husband.

Ever your affectionate Brother,

C. J. F. BUNBURY.

Interlaken,
September 10th, 1848.

My Dear Mr. Horner,

I thank you very much for your letter of the 20th of last month, containing a most interesting account of the scientific meeting at Swansea, and of the company you met there. I am delighted that you found it so agreeable and that so much good work was done,—and glad that you thought my information about the fossil Ferns worth reading to the *savants*, though, if I had foreseen that it was to have that honour, I would have entered into more details.

We have had a most delightful tour from Geneva, and, more fortunate than you in England, have enjoyed glorious weather ; indeed they say it is

1848. many years since Switzerland has had so fine a summer and autumn.

We stayed two days at Berne, where there is a pretty little museum, not comparable to those of Florence or Turin, but good as a local collection, rich in the animals of Switzerland and in the fossil shells of the *mollasse*.

These latter, unfortunately, I had not knowledge enough to appreciate, and I saw no fossil plants of any importance. The animals are remarkably well stuffed and mounted. Berne is a picturesque town, though not equal in that respect to Fribourg : it has a handsome Gothic cathedral, with a very curious sculptured porch, and some of its Gothic fountains are quaint and pretty; but its great charm is the splendid view of the snowy peaks of the Alps, which is displayed before one from its public walks, and which the weather fortunately allowed us to enjoy in full perfection. On Monday the 4th, a lovely day, we drove through the pleasant valley of the Aar to Thun, and thence along the south shore of the beautiful lake of that name, to this place, where we stayed the next day.

Interlaken is delightfully situated in a smooth green plain between the lakes of Thun and Brienz, with high and steep mountains on each side of it at very short distance, and the Jungfrau in all her glory seen through an opening between these mountains. I have her before my eyes at this moment, with the sun shining on her brilliant veil of snow. The little plain is occupied by bright green meadows, magnificent walnut trees, cheerful cottages of the picturesque Swiss fashion and (less picturesque, but very convenient), a dozen or so of hotels and lodging-houses. We are in a very comfortable hotel with a nice garden. On the 6th, we set out early in the morning for Lauterbrunen in a light carriage, which we left there, and crossed over the Wengern Alp to Grindelwald, an expedition of ten-and-a-half hours altogether, including halts. This was one of the most interesting and enjoyable excursions I ever made. The weather was delicious and I do not believe that any scene on earth can be more glorious than the view from the halting place on the Wengern Alp, directly facing the Jungfrau, with her sublime precipices, vast glaciers and dazzling fields of snow, and the two mountains, the Mönch and Eiger, which seemed but as her attendants, though anywhere else they would have been of the very first class. There is no describing such scenery; the impression which it makes can never be effaced nor forgotten, but all language seems faint and inadequate. We were fortunate in every way, for not less than ten or twelve avalanches fell down the cliffs of the Jungfrau while we were opposite to it, and we saw and heard them in perfection. The crashing roar of the avalanche, echoed and redoubled by the mountains, is exceedingly grand; but what one sees, appears hardly proportioned to such a sound; it looks exactly like a temporary cascade of snow pouring down the cliffs; the fact is that one's eye is deceived as to the distance, for the vast scale of the scenery and the absence of intervening objects make the

1848. opposite mountain appear quite close, though in
reality the distance is considerable, and what looks
like snow, really consists of great fragments of ice.
—The following day we crossed the Scheideck
to Meyringen in the valley of Hasli. This is
another very fine Alpine Pass, though we both
thought it considerably inferior to the Wengern Alp.
In the course of the day we visited two very
fine glaciers, the upper glacier of Grindelwald and
that of Rosenlaui, and walked some little way on the
ice under careful and experienced guidance. These
"ice palaces," which Forbes has so well described,
are very wonderful objects; when seen at a distance
they disappoint one, but when one is on them
or close to them, the vast rocks and ridges and
pinnacles of ice towering high over one's head, the
strange fantastic pendants and bridges of ice formed
by the water undermining the masses,—the deep
chasms, the beautiful transparent tints of sea-green
and blue, and the torrents rushing underneath or
falling in cascades through the crevices,—altogether
seem more like magic than reality. Milton's expression of "thick-ribbed ice" is brought forcibly
to one's mind in viewing them. The glacier of
Rosenlaui is famous above most others in Switzerland for the purity of its ice and the brilliant blue
colour in its chasms,—which is indeed exquisitely
beautiful. The old guide at the Grindelwald glacier
told us that fourteen years ago it had been much
further in advance than at present; that it had re-
ceded continually for seven years, then again
advanced, but in a much less proportion for seven

years, and that within the last year it had begun 1848. again to retire. He showed us the former *moraine* of the glacier, and pointed out one enormous block of stone in particular, which within his knowledge had been on the surface of the ice, and is now many yards in advance of it. I am not at all surprised at Agassiz's becoming quite enamoured of the glaciers and willing to see traces of their action everywhere, for there certainly is something very fascinating about them. Fanny was so much fatigued by these two days' work that I did not think it right for her to attempt the Grimsel, and so we returned by the Lake of Briénz to Interlaken.

(Berne, September 11th). You ask about the political state of Switzerland. It is rather difficult in so short a visit to the country to form any satisfactory opinion on such a subject,—impossible indeed to see more than the surface ; all I can say is, that to all appearance the country is in a state of profound tranquility and peace, and likely to continue so ; all traces of Revolution and civil war have disappeared as completely as if they had never been, and I do not suppose there is anywhere, at present, a more quiet country than Switzerland. In all the part of it that we have hitherto traversed, the appearance of general prosperity and comfort is striking ; it is a pleasure to travel through a country where the population seems so generally comfortable, industrious and thriving. But I must say that this appearance of material prosperity and plenty was equally striking when I was first in Switzerland, in

1848. 1829, when the old aristocratic governments were in full force; so whether the Swiss have gained anything by the Revolutions is not apparent on the surface.

I suppose you have read Mr. Grote's clever little book on Swiss politics? it is extremely well written, but somewhat in a partizan spirit, and the correctness of some of his facts is strongly disputed.

All our friends at Geneva, were of the aristocratic party, but I perceived no bitterness of feeling in any of them, except the De la Rives, who are violent Tories.

Speaking of Italian affairs, you say that a right feeling was probably wanting in the great mass of the people. I conceive that the masses in Lombardy cared very little about the matter, for the tyranny of Austria never weighed much on *them*, but was felt chiefly by the educated classes. The Government of Austria has kept the peasantry and mechanics in a state of considerable material comfort, while it has constantly and systematically striven to repress and extinguish all activity of intellect and all energy of character; in short its object has been to keep its subjects, as far as possible, in a state of contented submissive ignorance, and it has exerted a most vigilant tyranny in crushing all those more active and daring spirits that would not submit to be thus cramped. It has been so far successful that the listlessness and imbecility shown by the Milanese in this war, must be attributed in great measure to the effects of the Metternichian system. But Mazzini and his republican gang, have also materially served the cause of Austria.

I am very sorry that you will be in the land of 1848. smoke, and not at Rivermede, when we arrive in England; but I trust we shall meet before the winter is over. Much love to Mrs. Horner and our sisters.

Ever your affectionate Son-in-law,

C. J. F. BUNBURY.

Lucerne,
September 24th, 1848.

My Dear Mrs. Horner,

I wrote to Mr. Horner from Interlaken about a fortnight ago, giving him an account of our delightful excursion in the Oberland.

On the 21st we set out, together with Edward, who had arrived on the 15th, to see the upper part of the Lake, and the valley of the Reuss up to the St. Gothard; and we made a very enjoyable expedition of it. We went in the steamer to Flüelen, at the head of the Lake; hired a light carriage there, and drove to Amsteg, where we slept that night; the next day we went up the St. Gothard pass, as far as Andermatt in the Urseren Thal, dined there, and returned to sleep at Amsteg; and yesterday we took a row-boat from Flüelen to Brunnen, landed there, and returned to this place by way of Schwytz, Löwertz, Goldau, Arth, and Küssnach. This was a charming tour. The lake of the four cantons is very beautiful, much the most beautiful of all the Swiss Lakes, and it is rendered still more interesting by the great deeds and glorious names with which

1848. it is associated. There is scarcely any history more spirit-stirring than the early history of Switzerland, and the scenery of the three primitive cantons appears peculiarly in harmony with it, as if marked out by nature, for the scene of noble deeds. We landed at Tell's chapel, erected on the spot where he sprang ashore from the boat; and on the opposite shore we saw the green meadow of Grütli or Rütli, where the three famous men of the three cantons met to prepare the freedom of their country.

A little way from the head of the lake, in the village of Altdorf, are two fountains, one of which is said to be erected where William Tell stood to take aim at *the apple*, and the other on the site of the tree, to which his son was bound. We saw too, the spire of the village of Bürglen, his birth place, and the torrent (the Schächen) in which he perished. Altdorf though the chief place of canton Uri, is a poor decayed looking village, and all through that canton, we were struck with a general appearance of poverty and slovenliness beyond what we have seen anywhere else in Switzerland; beggars abound as in Italy, and the villages are very dirty. Uri is indeed the poorest and most barren of all the cantons; its whole territory consists of savage mountains and narrow Alpine valleys, frequently ravaged by torrents. Part of the St. Gothard Pass, between Göschenen and Andermatt, is very grand and striking scenery; the mountains rise on either side in tremendous walls of perfectly bare rock, leaving only space between them for a most furious foaming torrent, 1848. over which the road seems suspended.

(Zurich, September 28th). After I had begun this letter, Fanny and I made another interesting excursion from Lucerne, (Edward had parted from us at Schwytz, on the 23rd):—we went in a car to Winkel (I suppose you have the map before you), crossed over in a boat to Stanzsted, walked from thence to Stanz, with a man who carried our luggage ; dined at Stanz, the birth place of Arnold of Winkelried, whose statue stands in the market place ; hired a car there, and drove up a fine valley to Engelberg, where there is one of the greatest and most celebrated convents in Switzerland, situated in a beautifully wild secluded valley at a height of more than 3,000 feet above the sea level, and enclosed by grand mountains ; several of which are covered with eternal ice and snow. We slept there at a good inn, and the next morning visited the convent, and were shown the library of it by a very polite monk. Leaving Engelberg, we drove to Bekenried, on the shore of the lake of Lucerne, where we found a very nice inn, newly established ; there we dined and waited for the steamer, which took us back to Lucerne. The next morning (yesterday) we departed from Lucerne, with considerable regret, having made it our head-quarters for twelve days, and been very much pleased with all the neighbourhood. Lucerne itself is a remarkably picturesque town with a great number of quaint old towers, battlemented walls, and long old

1848. covered wooden bridges with curious paintings inside them. The views of the lake from it are lovely, and I do not know any place from which a greater variety of delightful excursions may be made. Well, we left it yesterday morning, and came to this place of which I can tell you nothing, as it has rained incessantly since we came. We shall set off tomorrow for Basle, and hope to arrive there the next day, and to find letters from both sides of the house. There we shall feel ourselves really drawing towards home, and I hope we shall make our way homewards without much delay. I have enjoyed Switzerland very much, and I think Fanny has too, but she is not as strong as she ought to be for the thorough enjoyment of it.

I am very sorry that you will not be at Rivermede when we return, but you must come and see us at Mildenhall.

Pray give my love to Mr. Horner, and to Joanna, who I understand is with you, and believe me,

Your very affectionate Son-in-law,
C. J. F. Bunbury.

Bonn,
October 18th, 1848.

My Dear Lyell,

We are now drawing pretty near to home and I hope to see you before long, when we shall have much to talk over. I suppose you will, in a very few days be settled again in Harley street, and if you keep to your plans you will be quite at home

again there before we reach Rivermede, which I am 1848.
afraid will not be earlier than the 28th, at soonest.

I was very much rejoiced to hear of the distinction
the Queen had conferred upon you—less for your own
sake, for after all a title can add but little to the real
distinction and honour which you have so well earned
for yourself — than because it is very pleasant to
see that our Queen and her Husband are able to
appreciate scientific merit, and to acknowledge its
value. I am glad of every example which tends to
counteract the prevailing English disposition to
undervalue all merit that is not political or military,
and although all who know you, must feel that you
deserve far higher honours than this, certainly the
way in which it has been done adds much to the
distinction, and makes it as gratifying as anything of
the kind can be. I am very glad to hear that
Prince Albert is so intelligent, and so well disposed
to encourage science ; he may do, and I hope will
do, a great deal of good.

I was very much pleased with the collection at
Strasbourg, and delighted with Professor Schimper ; I
introduced myself to him, and he showed me as much
attention and civility as if I had had a dozen letters of
introduction. Fanny was as much charmed with
him as I was ; he is indeed the most captivating
savant that we have met with in the course of our
travels ; full of ardour and enthusiasm, well informed
on a variety of subjects, and combining the agreeable manners of a Frenchman with the depth and
solidity of a German. You may perhaps have seen
his excellent book on the fossil plants of the Grés

1848. bigarré, which is in our Geological Society library. He is now engaged in a work on the fossil plants of the equivocal and ill-understood deposits of Scania, which he refers to the *Keuper*; he showed me several interesting specimens from thence, and some beautiful plates which he has prepared. But though a great authority on fossil botany, his special passion is for mosses, of which he knows more than almost any man, and he is the author of the most beautiful work that has appeared on European mosses; he has travelled over almost the whole of Europe, and ascended nearly all the mountains of this continent, and once in Lapland he undertook an expedition of 100 leagues to gather a particular rare moss. I was quite surprised by the riches of the Strasbourg museum, especially in fossil plants; those of the Grés bigarré, from the quarries of Soultz-les-bains are particularly fine and interesting, many of the specimens of Ferns extremely beautiful. It seems to be a peculiar and local deposit of fossils of limited extent, for M. Schimper told me that for the last five years he had not been able to obtain a single specimen from those quarries, and I cannot learn that anything similar has been found in Grés bigarré elsewhere. Another thing that interested me was a fine series of specimens of Equisetum columnare, from the *Keuper* sandstone of Wurtemberg, — precisely agreeing with the plant you procured from Richmond in Virginia, and which I had before been able to compare only with the specimens from the Inferior Oolite of Whitby. You would be pleased too with the magnificent specimen of fossil footsteps 1848. (Cheirotherium) in Grés bigarré from Hild- burghausen; and in short I think it would be well worth while to go to Strasbourg only to see the museum. At Heidelberg I saw Professor Leonhard, and bought several specimens of fossil plants from the Keuper and the Gres bigarré, in particular, a good authentic specimen of Calamites arenaceus, which I wanted to compare with the plant from your Virginian coal-field. There is at Heidelberg an establishment for the sale of fossils and minerals, under the management of a society, which M. Leonhard is anxious should be known in England, and he begged me to insert a notice of it in our " Quarterly Journal."

Heidelberg pleased me more than any other place I have seen in Germany,—indeed since we left Lucerne it is the only place with which I have been much captivated; but indeed it was the last place we were able to see satisfactorily, for ever since we left it the weather has been so perseveringly detestable that we have had no enjoyment in travelling. Since we quitted England we have met with no such continuance of bad weather as since the 9th of this month; and I feel that I cannot judge at all fairly of the famous scenery of the Rhine, for we made the voyage from Mayence to Bonn on a miserable cold, raw, dull, gloomy day, which put one out of sorts with everything. My impression is, that the scenery of the Rhone between Vienne and Avignon is superior to that of the Rhine, but it is hardly fair to form any judgment

1848. of the latter, under such unfavourable circumstances.

(*Aix la Chapelle, October 21st*).   Here we are, three days since the beginning of my letter, and the weather still as abominable as ever.   It is quite sickening, and makes me long to be quiet at home. In consequence of this, we stayed only a day-and-a-half at Bonn, as it was impossible to make excursions to Godesberg or anywhere; but we saw some of Fanny's friends: and I examined the museum, with which I was much pleased.   It was shown to me by a pleasing young man, Dr. Roemer, who desired to be particularly remembered to you, and spoke of your kindness and attention to him in London, before he went to America.   He travelled much in Texas, and made great collections there, which are however not yet arranged or exhibited.   In the museums, some of the things which struck me most were, — first of all, the unique specimen of a Saurian from the coal formation of Saarbrück; secondly an enormous *Mosasaurus* from the Missouri, which must surely be the original sea-serpent, its vertebral column is nearly complete, and must be near 40 feet long, then the fossil frogs and insects in the brown coal, the crustacea and insects from Solenhofen, and the Encrinites and other fossils of the Muschelkalk. The fossil plants did not appear to me particularly interesting.   Bonn may be a charming place in fine weather, but certainly it did not appear to me under a fascinating aspect.   However, it still looked much better than Cologne, where we spent yesterday and this morning. The cathedral at Cologne would make up for any disagreeables; it is the most glorious work of human hands that I ever beheld—far superior (to my mind) to St. Peter's; and they are continuing it in the most admirable style, the new work being quite equal to the old. But I am afraid that these odious political tumults and troubles in Germany, will for a long time interrupt, if they do not entirely prevent its completion.

Most persons with whom we have talked, think very ill of the political prospects of Germany; our friends at Bonn talked very despondingly.

M. Arnott alone (whom we saw at Frankfort) seemed hopeful, yet he would not enter into any particulars and rather appeared anxious to escape from the subject.

M. Schimper prophesies that Germany will have to pass through a trial as terrible as France underwent in the first Revolution; and really these horrible murders at Vienna look very like it.

All the way down the Rhine, ever since we left Basle we have found the towns swarming with soldiers, and they are very necessary. We seem further than ever from that era of universal peace and tranquillity, which was so much talked of last winter.

I now hope to get to Rivermede by the 26th or 27th, as we learn that the railway is now open

1848. from Ghent to Calais, and that that journey
may be made in six hours. I shall be very happy
to see you and dear Mary again. Pray give
my love to her, and believe me

Ever your affectionate Brother-in-law,

C. J. F. BUNBURY.

MILDENHALL :
PRINTED BY S. R. SIMPSON, MILL STREET.